91 **Topics in Current Chemistry**
Fortschritte der Chemischen Forschung

Syntheses of Natural Products

Springer-Verlag
Berlin Heidelberg New York 1980

This series presents critical reviews of the present position and future trends in modern chemical research. It is addressed to all research and industrial chemists who wish to keep abreast of advances in their subject.

As a rule, contributions are specially commissioned. The editors and publishers will, however, always be pleased to receive suggestions and supplementary information. Papers are accepted for "Topics in Current Chemistry" in English.

ISBN 3-540-09827-5 Springer-Verlag Berlin Heidelberg New York
ISBN 0-387-09827-5 Springer-Verlag New York Heidelberg Berlin

Library of Congress Cataloging in Publication Data. Main entry under title: Syntheses of natural products. (Topics in current chemistry ; v. 91) Bibliography: p. Includes index. Contents: Warren, S. G. Reagents for natural product based on the Ph_2PO and PhS groups. – Tsuji, J. Application to natural products syntheses. – Schuda, P. F. Aflatoxin chemistry and syntheses. 1. Natural products – Addresses, essays, lectures. 2. Chemistry, Organic – Synthesis – Addresses, essays, lectures. I. Warren, Stuart G. Reagents for natural product based on the Ph_2PO and PhS groups. 1980. II. Tsuji, Jiro, 1927 – Application to natural product syntheses. 1980. III. Schuda, Paul F., 1952 – Aflatoxin chemistry and syntheses. 1980. IV. Series. QD1.F58. vol. 91. [QD415.2]. 540s. [547.7'0459] 80-12663

Typesetting and printing: Schwetzinger Verlagsdruckerei, 6830 Schwetzingen
Bookbinding: Konrad Triltsch, Graphischer Betrieb, 8700 Würzburg
2152/3140–543210

Contents

Reagents for Natural Product Synthesis Based on the Ph_2PO and PhS Groups

Stuart G. Warren

University Chemical Laboratory, Lensfield Road, Cambridge CB2 1EW, England

Table of Contents

1 Introduction

The disconnection approach to the design of organic syntheses was devised by Corey[1] largely for use by computers[2]. It is equally suitable for use by people[3]; even Corey has discarded his IBM on occasions[4]. In some respects this fundamental approach contrasts with the "key reaction" approach used, for example, by Oppolzer[5] in his recent synthesis of Longifolene[a]. However, both approaches have focussed attention on synthons[b] – idealised fragments into which the target molecule may be broken, and on the need for reagents – corresponding to these synthons which are regiospecific in the sense that they have nucleophilic or electrophilic reactivity at one specific carbon atom[9].

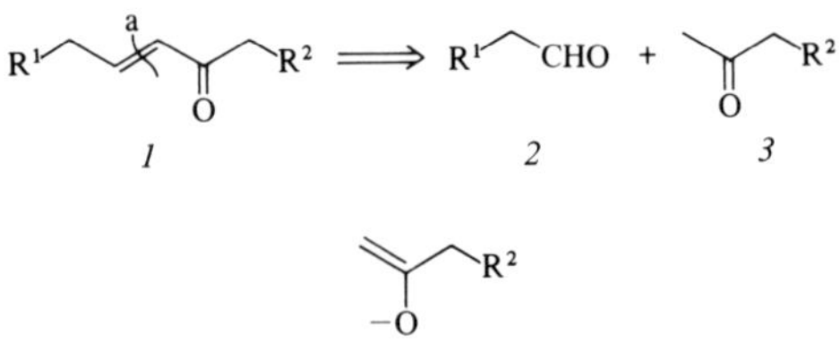

1 *2* *3* *4*

Conventional disconnection of the enone *1* at *a* suggests an unlikely aldol condensation between the aldehyde *2* and the unsymmetrica ketone *3*; unlikely because one specific enolate *4* out of a possible three must react with only one of the two carbonyl groups. Successful solutions to this problem may use specific enol equivalents[10] for *4* such as enamines[11] or enol silanes[12]; both have been used in the synthesis of many natural products. Alternatively, we may choose a different disconnection, say *b* in *5*, and use an epoxide and the acyl anion *6*, a synthon with inverted polarity or umpolung[9, 13]. Reagents for *6* include those based on sulphur[14], particularly dithians[15], and this strategy has been used for the synthesis of a few natural products, mostly enals[16].

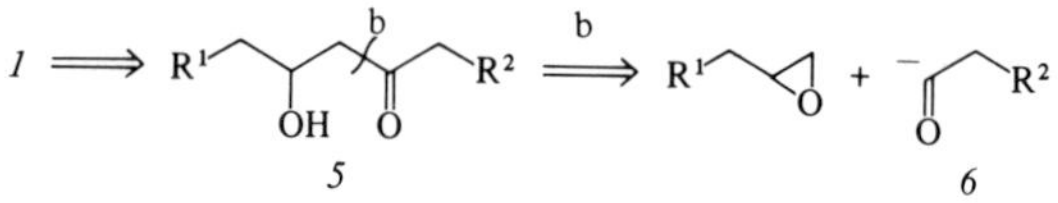

5 *6*

Specific enol equivalents (e. g. β-keto esters) and umpolung (e. g., with cyanide or acetylide ions as acyl anion equivalents) have of course been used in synthesis for many years. What is new is the recognition of their role, as a result of the disconnec-

---

[a] A compound previously regarded as the province of the disconnection approach[6]

[b] *Synthon* was originally used by Corey[7] to mean a part structure within the target molecule recognisable as the product of a known reaction. He now suggests retron for this meaning[8]. Synthon is often loosely used to mean a reagent but the most helpful usage[3, 9] is *synthon* for the idealised fragment (e. g. R^- and *reagent* for its equivalent (e. g. RLi) in the flask.

tion approach, and it is this that has led to the many recent developments. The same is true of functional group transpositions[17, 18]. Alkylative carbonyl transpositions[18] have proved particularly useful in the synthesis of natural products whose complicated structures are often much easier to analyse after disconnecting a C–C bond and relocating a carbonyl group at a more convenient site in a single operation.

Examples of two types of 1,2-alkylative carbonyl transpositions occur in Corey's occidentalol synthesis[19], and in Oppolzer's acorenone synthesis[20]. In the first, *7→8*, the carbonyl group is moved forwards into the newly added part of the molecule. In the second, *9 → 10*, the carbonyl group is moved backwards around the original framework.

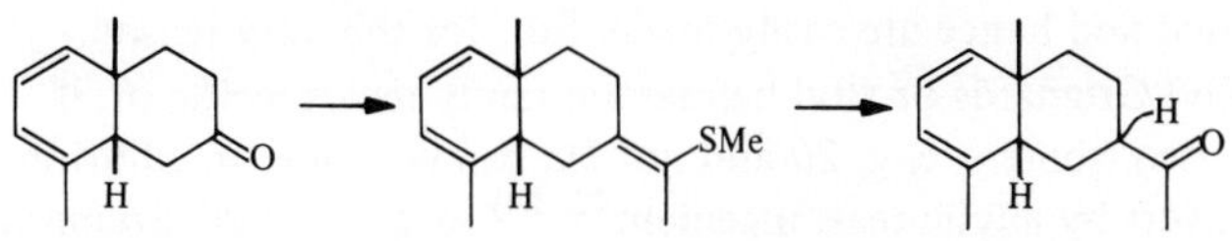

7 *8*

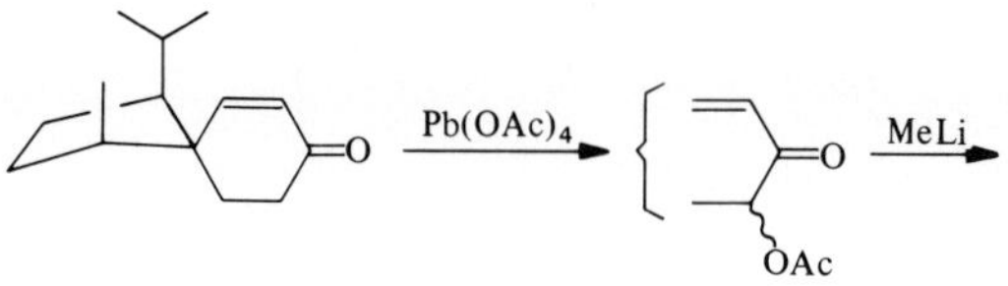

9

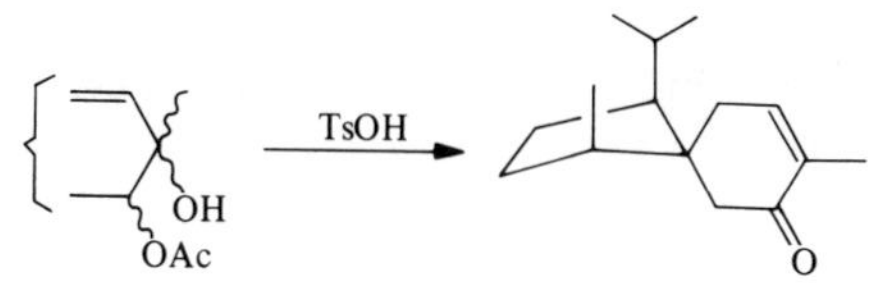

10

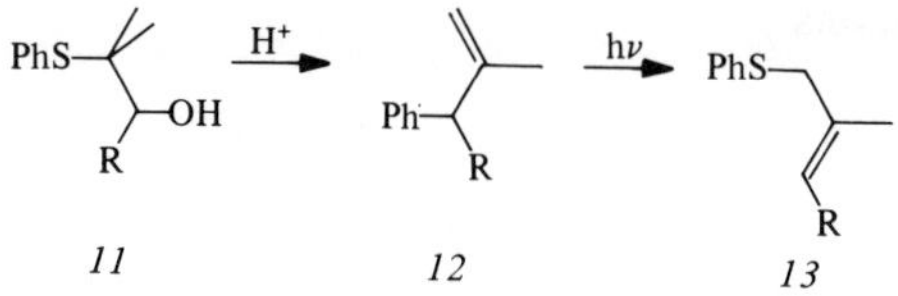

11 *12* *13*

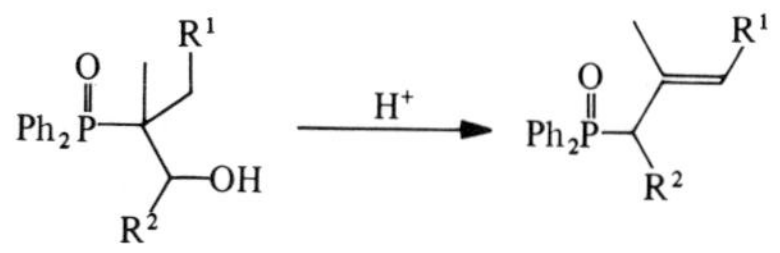

14 *15*

We have found[21] that two functional groups, phenylthio (PhS) and diphenylphosphinoyl (Ph_2PO) will take part in a slightly different operation – migration – and form the allyl derivatives *12* and *15* in the acid catalysed rearrangement of alcohols *11* and *14*. The allyl sulphide *12* also rearranged photochemically by a [1,3] PhS shift[22] to the new allyl sulphide *13*. Since both Ph_2PO and to a lesser extent

PhS stabilise carbanions, these allyl compounds *12, 13,* and *15* form anions which can be used in synthesis. This article describes further developments in our work with these two groups, involving both migration and transposition, leading to regiospecific reagents for synthons with normal reactivity or umpolung and their application to natural product synthesis.

2 Regiospecific Allyl Synthons

Allyl anions and cations, e. g. *18,* are useful in synthesis[23] in that they are stable because they are delocalised and hence are easily made. But, for this very reason, simple reagents such as allyl Grignards or allyl halides are rarely regiospecific or, if they are, only one of the two isomers, e. g. *20* and not *19,* can be made. In addition, the reagents may interconvert by allylic rearrangement[22, 24], e. g. *19* ⇌ *20.* So many natural products contain allyl groups, particularly allylic alcohols, that the need for regiospecific reagents for allyl anions and cations is very great.

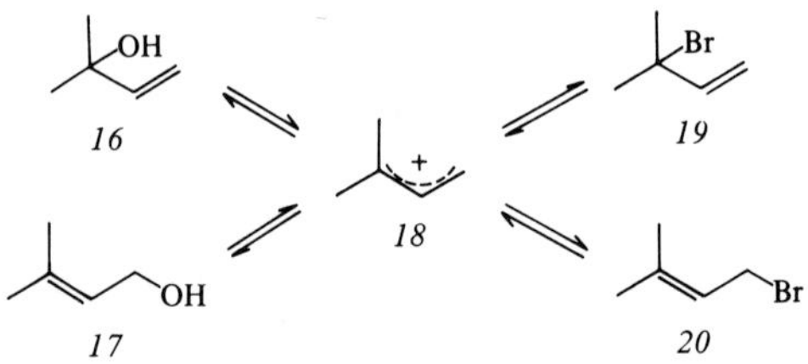

2.1 Regiospecific Reagents for Allyl Anions

Though some progress has been made in understanding the regiochemistry of allyl anions[17, 25], their behaviour towards nucleophiles remains capricious. The anions of allyl diphenylphosphine oxides are one of the most straightforward in their reactions with carbonyl compounds. The allyl phosphine oxide *23,* made by direct dehydration of *22* or by Ph_2PO migration from *21,* gives[26] an anion with butyl-lithium (BuLi) which reacts exclusively α to phosphorus with aldehydes to give the alcohols *24.* Each diastereoisomer of *24* (they are easily separted by chromatography) gives a single geometrical isomer of the diene *25.*

A degree of control over the geometry of the other double bond is also possible[27]. Dehydration of either alcohol *26* or *27* gives mostly the *E*-allyl phosphine oxide *28* and this configuration is retained throughout anion formation, addition to give *29,* separation into diastereoisomers, and elimination to give a single geometrical isomer of *30.* One double bond in the diene *30* must have the *E* configuration, but the configuration of the other is determined by which isomer of *29* is used.

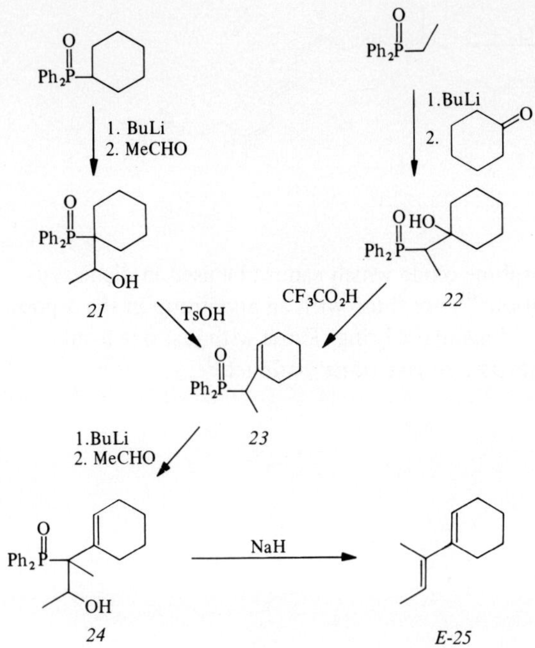

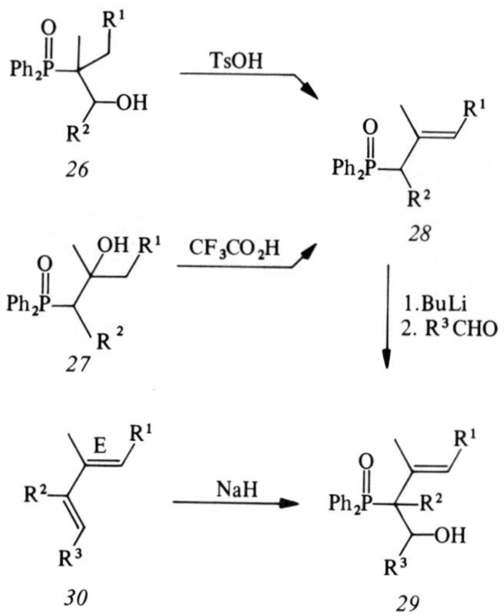

The allyl phosphine oxide approach to diene and polyene synthesis has been used by Lythgoe[28] in his vitamin D_3 synthesis, and by Pattenden[29] in syntheses of compounds such as *31*.

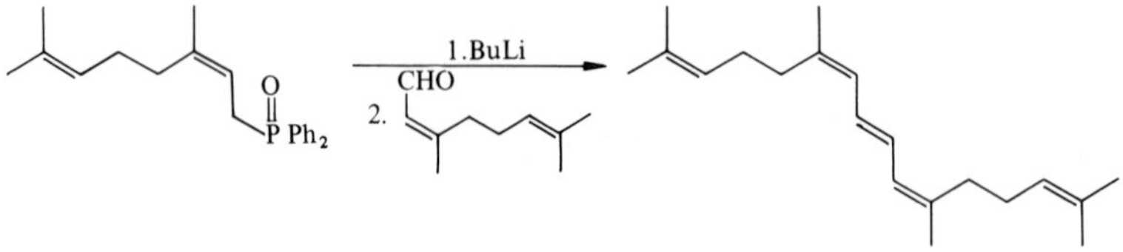

31

The main classes of allyl phosphine oxide which cannot be used in diene syntheses, as they react in the γ-position[30], are those with an aryl group in the α-position *32,* or with a four *33* or five *34* membered ring. Those with just one α-substituent and no β- or γ-substituents *35* also give some γ-adduct[31].

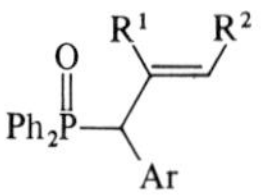

32

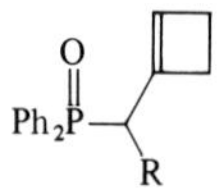

33

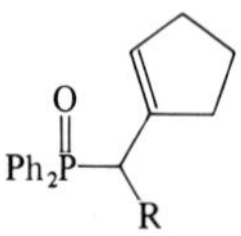

34

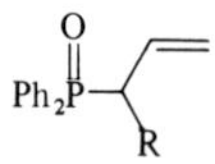

35

2.2 The Allyl to Vinyl Transposition: Synthesis of Allyl Alcohols from a Vinyl Anion Equivalent

The carbonyl adducts of allyl phosphine oxides, e. g., *24* and *29* can also be used to make allyl alcohols[32] by the reductive removal of the Ph_2PO group with lithium aluminium hydride: a reaction involving transposition of the double bond, e. g., *24* → *36.* In this approach, the allyl anion of *23* is a reagent for the vinyl anion synthon *37.*

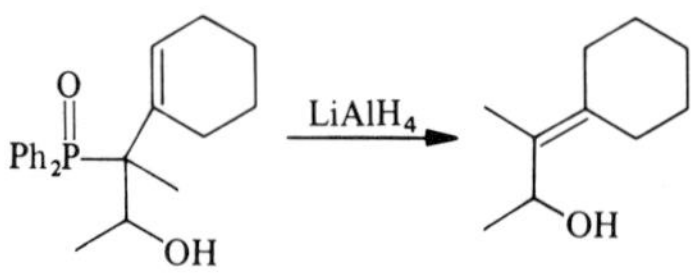

24 *36*

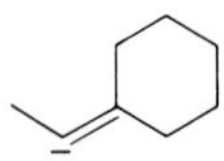

37

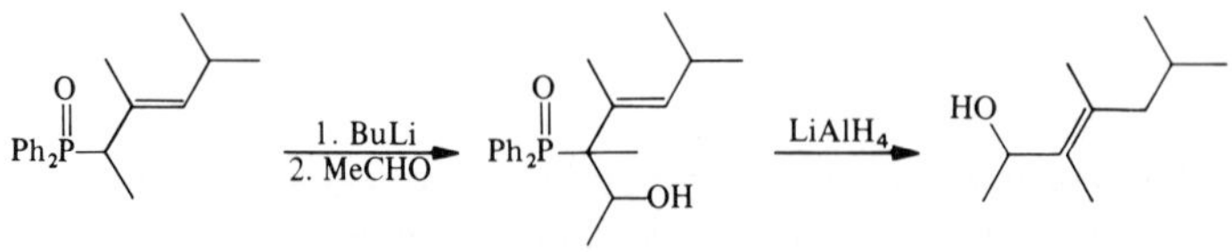

38

In open chain compounds, the *E* isomer of the allyl alcohol is formed stereoselectively, e. g., *38* in 80% yield[32].

2.3 Regiospecific Allyl Cation Reagents

One of the most widely used allyl alcohol syntheses uses the Evans-Mislow rearrangement (*39*, arrows) of allyl sulphoxides[17] (Scheme 1). Since the allyl sulphoxide is usually made from some allylic electrophile by substitution and oxidation, this strategy requires a specific allyl cation equivalent.

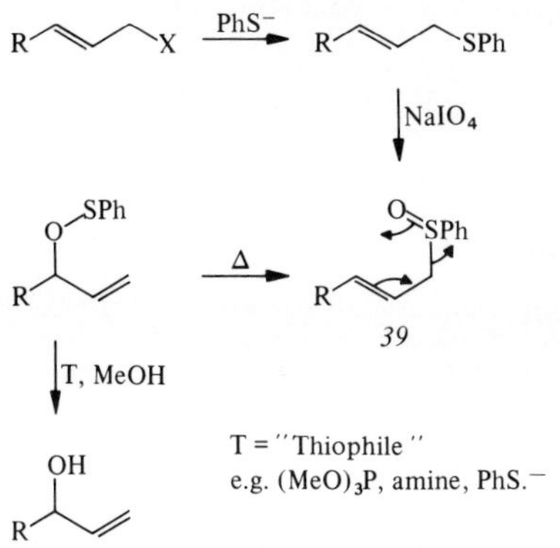

Scheme 1. Allyl alcohol synthesis *via* allyl sulphoxides

Evan's synthesis[17] of Yomogi alcohol *43* shows an ingenious solution. The sulphide *40*, available from bromide *20* as it contains only the more stable of the two allyl groups, is transformed by a [2,3] sigmatropic rearrangement (*41*, arrows) into a sulphide *42* with the *less* stable allyl fragment incorporated into its structure. Oxidation and reaction with the thiophile Et_2NH gives Yomogi alcohol *43*.

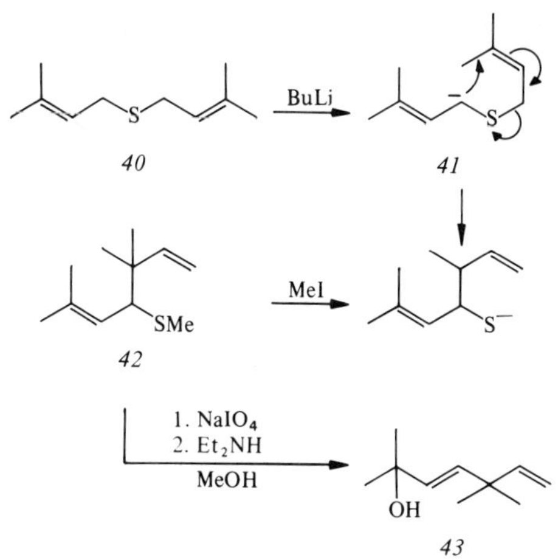

Our PhS migrations give an alternative solution[21, 33]. Acid catalysed rearrangement of alcohol *44* gives the allyl sulphide *45* and hence the allyl alcohol *47*. Photochemical rearrangement[22] of *45* gives the *more* stable allyl sulphide *46* and hence the *less* stable allyl alcohol *48*. The PhS group migrates very well and the scope of this reaction is wider than of Ph_2PO migration, allowing the synthesis of such allyl sulphides as *49*, *50*, and *51*. With the aid of silicon[34] the scope is even wider, so that the prenyl sulphides *52* and *53* can be made. Hence the original problem of making *16* and *17* regiospecifically is solved.

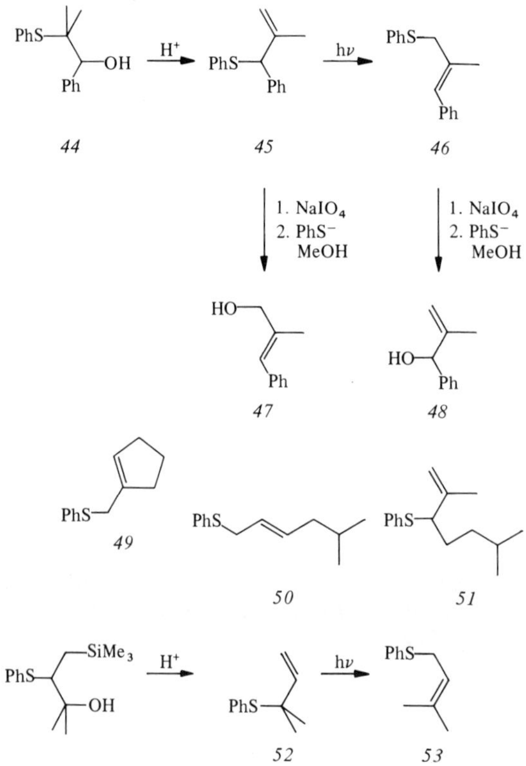

Analysis of synthetic routes involving functional group migration is sometimes so complex as to be unhelpful. The starting materials for the PhS migrations are usually made by using a sulphur-stabilised anion[33], e. g. from *54* in scheme 2, so that *54* behaves as the synthon *55*, or *56* if the [1,3] PhS shift is not used. These synthons are hardly likely to inspire organic chemists with enthusiasm and it is better to regard the rearrangements as key reactions. The allyl sulphides are conceptually simpler. Their anions may be alkylated at the sulphide or sulphoxide, e. g. *57* stage, and correspond to synthons such as *58*. This strategy was used by Evans[35] in his synthesis of a prostagladin intermediate. Comparison of synthons *58* and *37* shows how the two allyl alcohol syntheses differ.

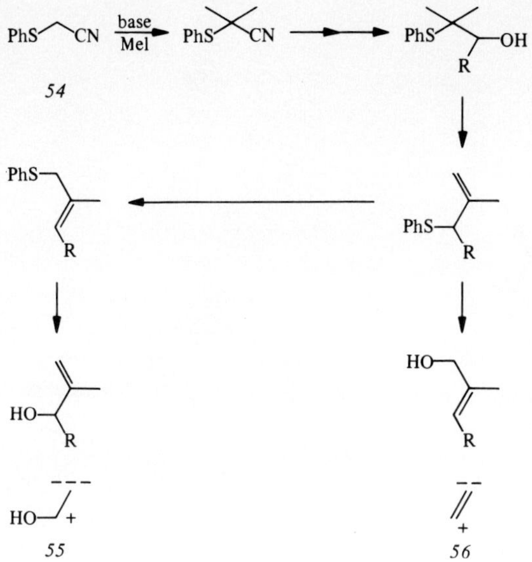

54 55 56

Scheme 2. Analysis of the allyl alcohol synthesis by PhS migration

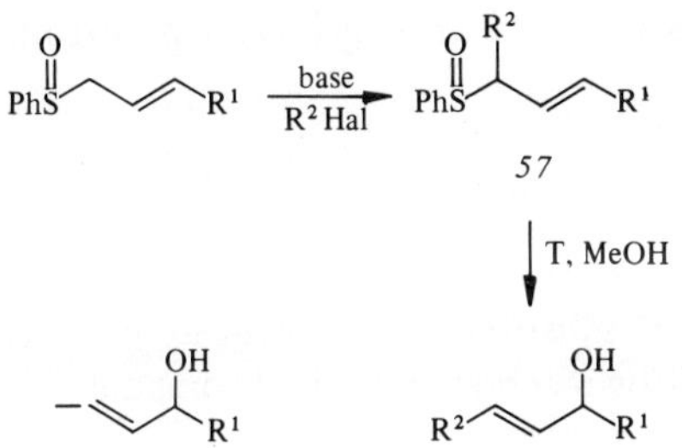

57 58

2.4 Epoxides of Allyl Phosphine Oxides as Allyl Cation Equivalents

The epoxides *59,* easily made from our allyl phosphine oxides *15* with metachloro perbenzoic acid (MCPBA) are clearly regiospecific allyl cation equivalents as nucleophiles attack the less substituted site giving alcohols, e. g. *60,* ready to complete the Horner-Wittig[26)] reaction. We have only started to explore the chemistry of these stable crystalline oxides *59* but one nucleophile which reacts cleanly[36)] is PhS^- giving the alcohols *60* and hence the allyl sulphides *61.* Completion of the sequence gives the allyl alcohols *62* so that the final result amounts to a direct displacement of the Ph_2PO group in *15* by water!

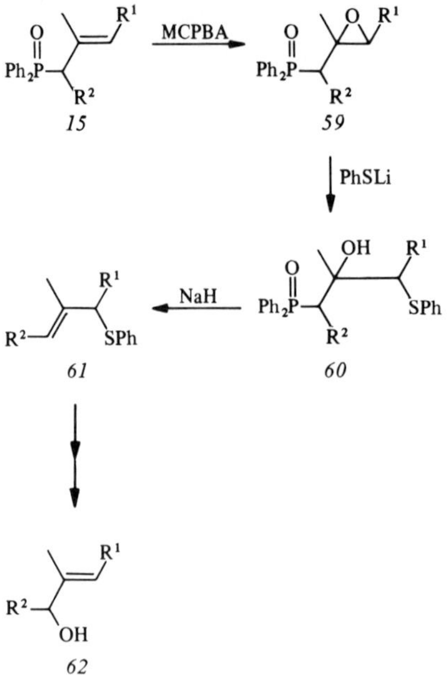

3 Acyl Anion Equivalents for 1,2-Alkylative Carbonyl Transpositions

In 1,2-alkylative carbonyl transpositions where the carbonyl group moves forward into the newly added fragement, e. g. *63* → *64*, the reagent behaves as an acyl anion equivalent. We have seen one example of this in the synthesis of the occidentalol intermediate *8*. Phosphine oxides with OR or SR substituents on the α-carbon *65* are ideal reagents for this process as the Horner-Wittig reaction gives vinyl compounds which can be hydrolysed to *64*.

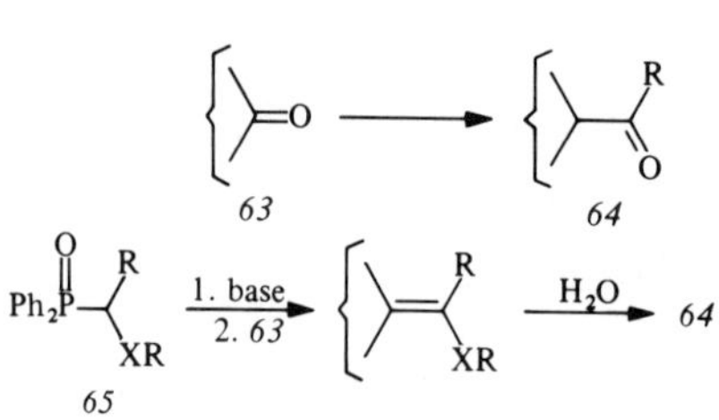

3.1 Synthesis of Aldehydes *63* → *64*, R=H

The classical solution to this problem is the decarboxylation of glycidic acids[37] *66*, used by Woodward[38] in his Lysergic acid synthesis. The Wittig reaction was modified by Wittig and Schlosser[39] using the ylid *67* to make vinyl ethers and hence the required aldehyde but this method has proved awkward in practice. Attempts to convert 20-oxo-pregnane into a bufadienolide by this route failed, though model reactions were fairly successful [40].

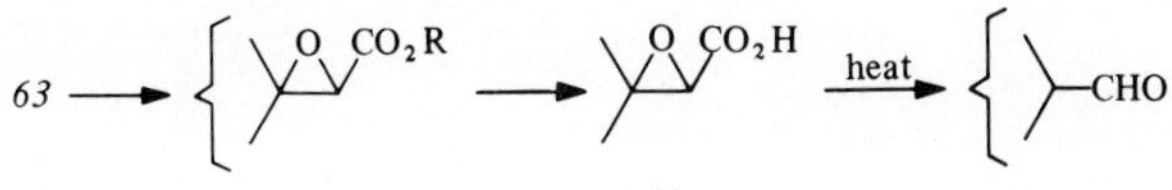

66

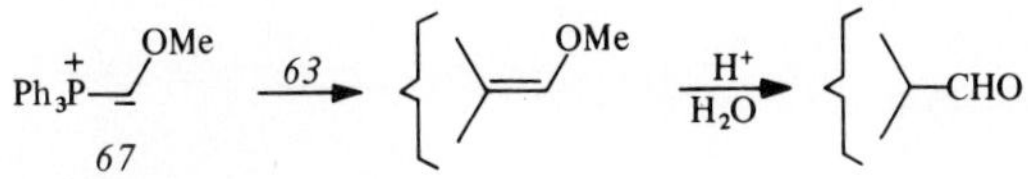

67

Phosphine oxide anions are often superior to ylids in olefination reactions, and the anion of *68*, made with lithium di-isopropyl-amide (LDA), has none of the disadvantages of the ylid *67*. We have made [41] a range of vinyl ethers *70* this way[c], and as part of a synthesis of strychnos alkaloids[43], we were able[41] to convert the acyl indole *71* into the aldehyde *72*.

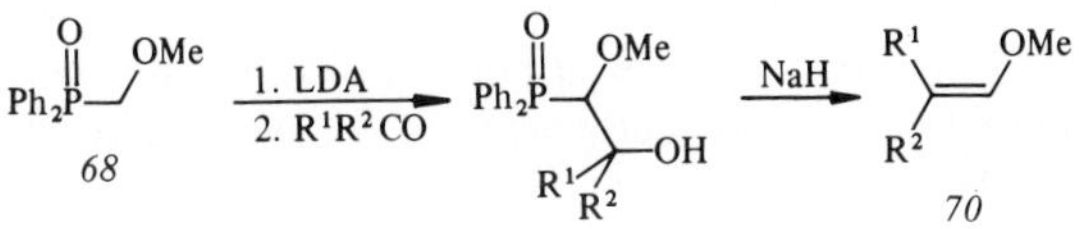

68 *69* *70*

71 *72*

One advantage of this route is that the vinyl ethers can be synthesised as single geometrical isomers by separating the diastereoisomers of *69* and converting each separately into the vinyl ether[d]. These vinyl ethers form anions which have been used in synthesis[45]. The trans vinyl ether *E-73* tends to form the vinyl anion[46] *74*, a reagent for the acyl anion synthon *75*, whilst *Z-73* tends to form the allyl anion[47] *76*, a reagent for the homoenolate synthon *77*. A reaction of this kind was used by Baldwin[46] to convert oestrone into the ketone *78*.

---

[c] Schlosser[42] reports briefly one example of the use of the same reagent

[d] Single geometrical isomers of vinyl ethers have been made by Hudrlik[44] from vinyl silanes

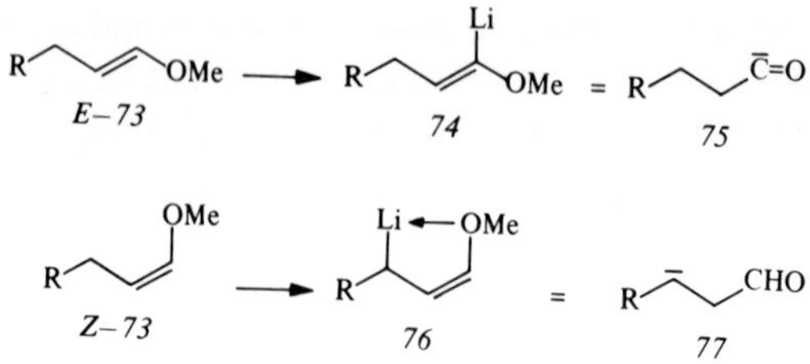

E–*73* *74* *75*

Z–*73* *76* *77*

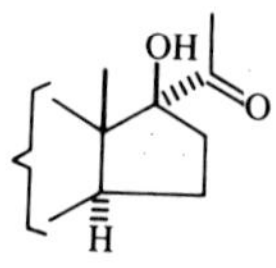

78

3.2 Synthesis of Ketones

3.2.1 From Sulphenylated Phosphine Oxides

Though *78* was synthesised *via* a vinyl ether, we find vinyl sulphides better intermediates in the synthesis of ketones by the transposition *63* → *64* because the reagents *79* are more readily available. They are easily made[48, 49] by direct sulphenylation of phosphine oxides and they form anions with BuLi which react cleanly with aldehydes, or with ketones if MeS is used in place of PhS[49], to give the vinyl sulphides *80* in one step. Dissolving *80* in trifluoroacetic acid (TFA) gives the ketones in high yield[48–50]. We made *81* in 93% yield from *79* (R^1 = CH_2Ph) and *82* in 86% yield from *79* (R^1 = Pr-i) by this route.

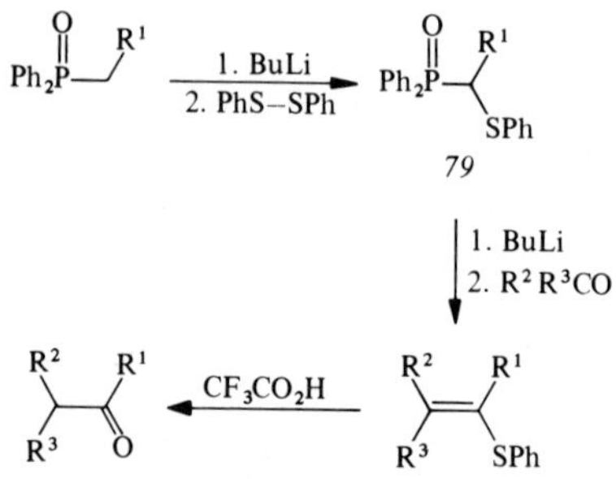

79 *80*

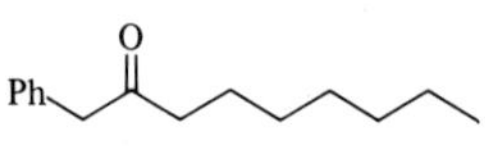

81

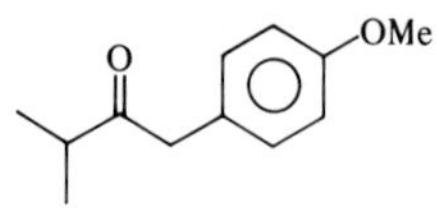

82

The corresponding phosphonium salts[51], and phosphonate esters[52] e.g. *83*, have also been used in ketone syntheses by this route. The phosphonate esters have been most widely used, especially in transposing cyclic ketones out into a new side chain, as in *7*→*8*. In Pinder's synthesis[53] of valencene and nootkatone *85*, the carbonyl group in *84* is not present in the final product *85*, but it is needed so that the side chain can be moved down to an equatorial position by base-catalysed epimerisation. Anions of vinyl sulphides have been used in a synthesis of dihydrojasmone[50].

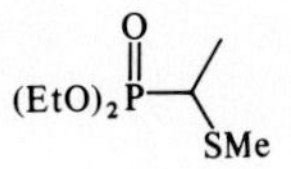

83

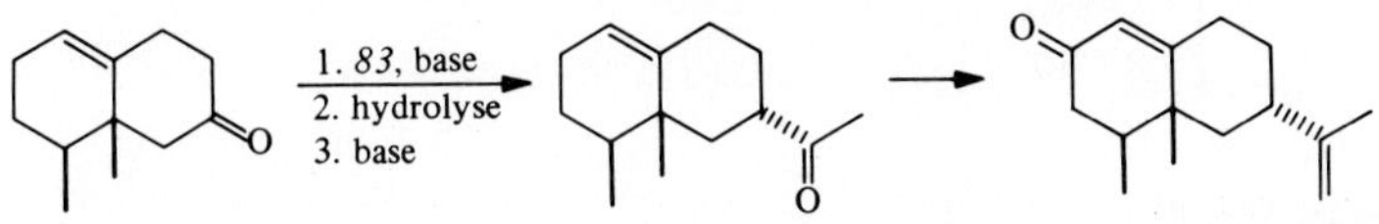

84 *85*

3.2.2 From Bis-(Phenylthio-) Acetals

The anions of *68* and *79* are then useful acyl anion equivalents, but in natural product syntheses they are far less popular than reagents containing two sulphur atoms[54], particularly dithians[55]. The chemistry of dithians, e. g. *86* and *87* has been well explored and they have been used in the synthesis of many natural products[e]. The synthesis[56] of the Douglas-Fir tussock moth sex pheromone *88* is an example of the way a dithian *86* acts as an acyl anion equivalent in ketone syntheses.

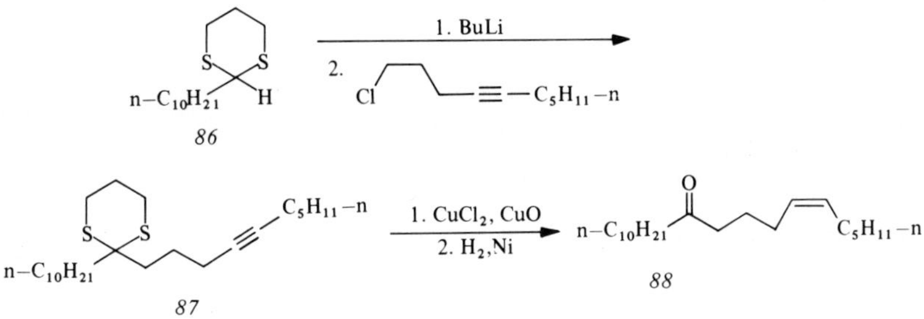

86 *87* *88*

For ketone synthesis we believe[48] there are distinct advantages in using the 1,2-alkylative carbonyl transposition *63* → *64* with bis-(phenylthio-) acetals *89*. The reagents *89* are made directly from the aldehyde or by alkylation of bis-(phenylthio)-methane. Anion formation with BuLi and addition to an aldehyde gives high yields of adducts *90* whereas the second alkylation of dithians with alkyl halides can give poor yields. The main advantage over dithians, however is in the remarkable reaction of these adducts *90* with TFA.

---

[e] There is a list in Ref.[54]

The required ketone *91* is formed almost intantaneously: the other product is diphenyl disulphide. We have some evidence[57)] for a mechanism involving PhS migration for this reaction. As examples, ketones *92* (64% yield) and *93* (88%) can be made this way[57)].

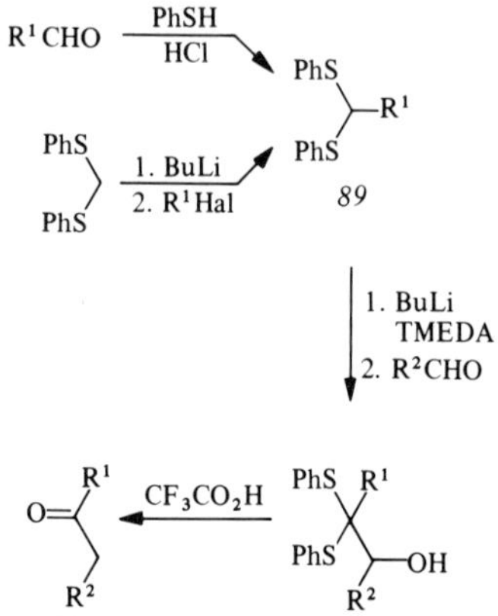

89 *90* *91*

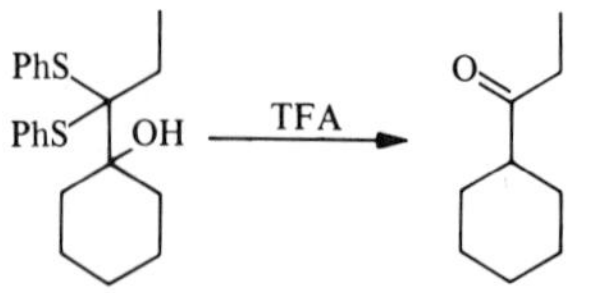

92

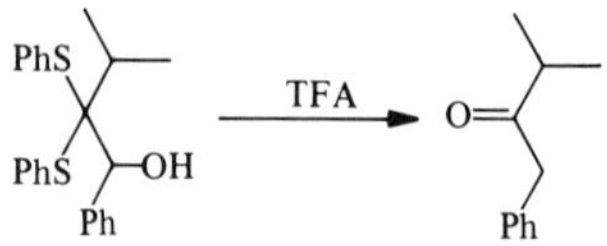

93

4 Specific Enol Equivalents

4.1 α-(Phenylthio-) Ketones

Acyl anions require special chemistry because they have umpolung of reactivity: specific enols have normal reactivity but the problem of regioselectivity must be solved. The carbonyl compound must enolise under the reaction conditions only on the required side and, as carbonyl compounds are also electrophilic, it must not condense with itself under these conditions.

α-(Phenylthio) ketones, e. g., *94,* fill this role admirably: the sulphur atom stabilises anions well enough for the enolate *95* to be formed with NaH or t-BuOK in tetrahydrofuran (THF) and to be stable under these conditions. The enolate *95* is still reactive enough to combine with alkyl halides[58)], as in Monteiro's synthesis[59)] of methyl dihydrojasmonate, or with vinyl sulphoxides, as in Schlessinger's synthesis[60)] of PGE, precursors.

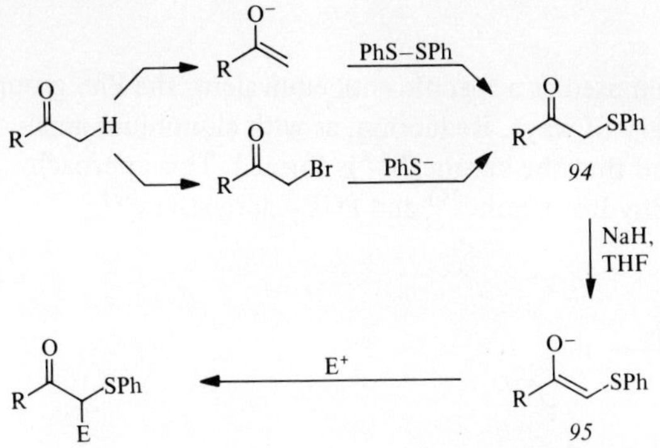

Unfortunately, the usual routes to α-PhS ketones *94*, from the parent ketone[33, 61, 62] by sulphenylation or halogenation, themselves require a solution of the specific enol problem! This can be found if silyl enol ethers are combined with PhSCl[63].

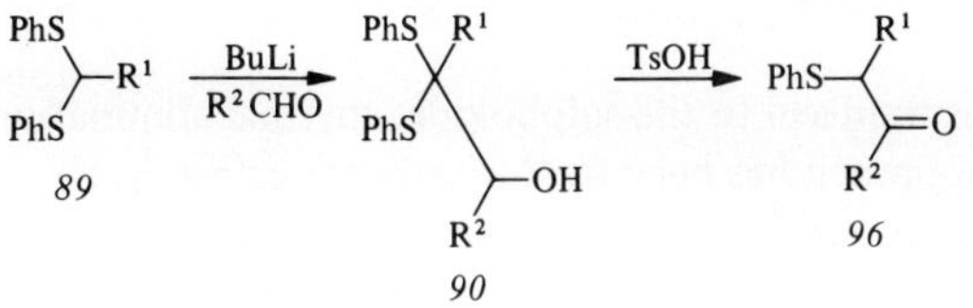

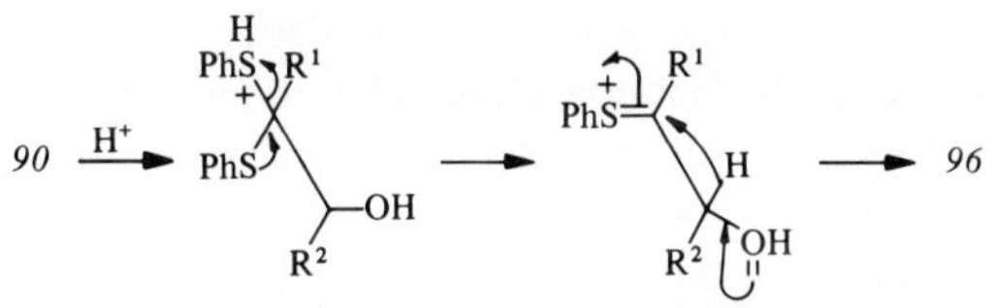

97

While we were working with the adducts *90*, we discovered[64] that treatment with TsOH gave α-PhS-ketones *96* directly. The reaction is at least partly[57] a hydride shift *97*. Regiospecificity is inevitable as *90* is made by joining the two halves of the molecule together so that both α-PhS regioisomers of any ketone, e. g., *98* and *99* or *100* and *101* can be made separately and unambiguously[57].

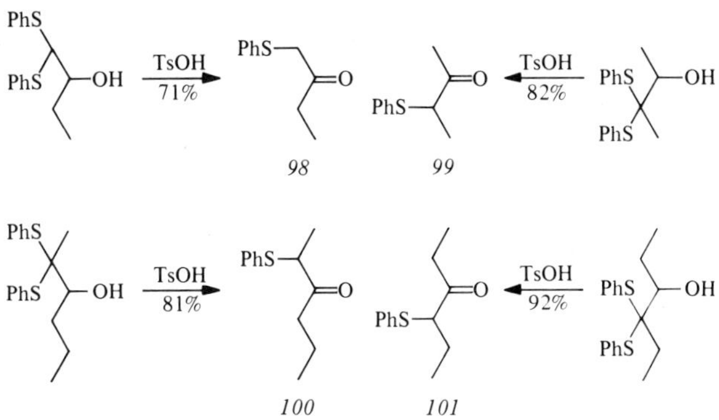

4.2 Synthesis of Butenolides

Once the α-PhS-ketone *102* has been used as a specific enol equivalent, the PhS group can be removed from *103* in a variety of ways. Reduction, as with aluminium amalgam[59], simply replaces PhS by H so that the ketone *104* is formed. This approach has been used in the synthesis of dihydrojasmone[59] and PGE_2 derivatives[65].

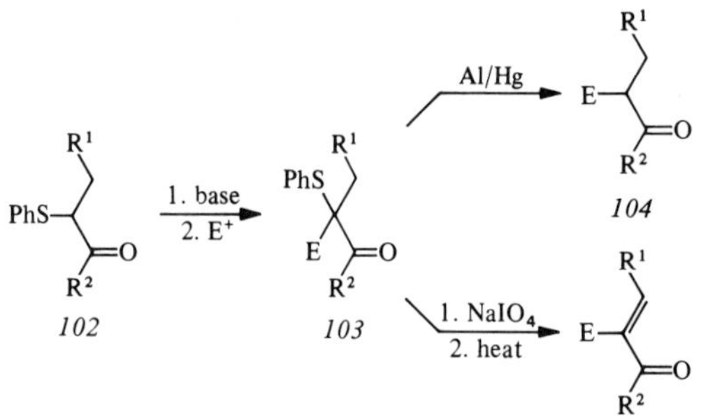

102 *103* *104* *105*

More commonly, the sulphide is oxidised to the sulphoxide: thermal elimination then gives[62] the enone[f] *97*. This approach has been used mostly for esters, as in Trost's synthesis of honey bee pheromones[62], or in syntheses of the important α-methylene lactones[66] including many natural products such as avenaciolide.

We have adapted[67] the approach to the synthesis of butenolides *108*. This group of compounds includes such natural products as the cardenolides[68]. The enolate of the α-PhS ketone *96* reacts cleanly with iodoacetate *anion* to give the keto-acid *106*. Reduction gives the lactones *107* (diastereoisomers), oxidation and elimination give the butenolide *108*. Since the sulphur atom is β to the carbonyl group in *107*, elimination occurs very easily and there is no ambiguity in the position of the double bond.

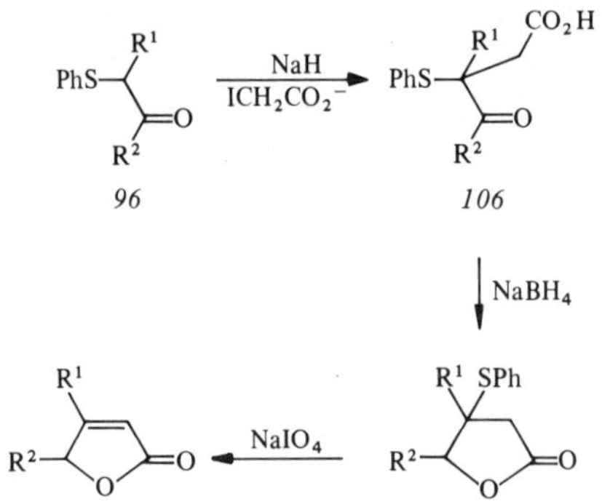

96 *106* *107* *108*

---

[f] In the more general enone synthesis *89* → *102* → *105*, the original anion of the bis-(phenylthio) acetal *89* behaves as a vinyl dianion ($R'CH{=}C^{2-}$), an interesting complement to synthons *37* and *58*

α-RS ketones can also be used in 'backwards' alkylative carbonyl transpositions, as in Trost's synthesis[69] of acorenone *10,* in epoxide syntheses[70], Beckmann fragmentations, as in Grieco's vernolepin work[71], and oxidations either to α-diketone derivatives or ring cleavage products, both effective on steroidal ketones[72].

5 Reagents for the Synthesis of 1,4-Dicarbonyl Compounds and Enones

This section is for two useful reagents, the allyl alcohol *111* and the enone *112.* Both can be made[73] by rearrangement of the adduct *102* which we have already used to make ketones and α-PhS ketones in the last two sections. On rearrangement with thionyl chloride and triethylamine (instead of acid) one PhS group migrates to give compounds *109* containing one allyl and one vinyl PhS group. Oxidation with

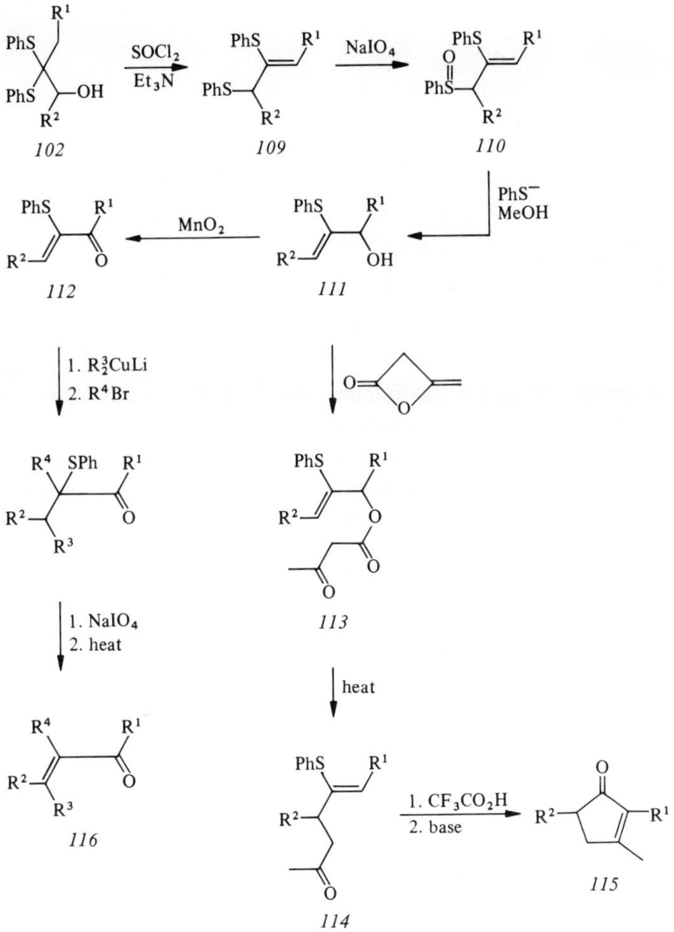

102 *109* *110* *112* *111* *113* *116* *114* *115*

sodium metaperiodate converts only the allyl PhS group to the sulphoxide *110* which gives the allyl alcohols *111* by the usual Evans-Mislow rearrangement[17]. The enones *112* are formed from *111* by oxidation with MnO_2[73]. Both *111* and *112* are formed as *E,Z*-mixtures but this is not important in the applications which follow.

The allyl alcohols *111* react with diketene to give compounds *113* which undergo the Carroll reaction on heating. The products *114* are half-masked 1,4-dicarbonyl compounds which give cyclopentenones on hydrolysis[50]. The enones *112* are good Michael acceptors and can be used to make enones *116* by addition of a nucleophilic and an electrophilic fragment followed by the usual sulphoxide elimination[59]. The enones *112* are therefore reagents for the synthon *117*. Dihydrojasmone has inevitably been made by both routes, though the intermediates *111* and *112* were made by other methods.

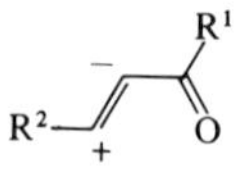

117

6 Homoenolate Equivalents

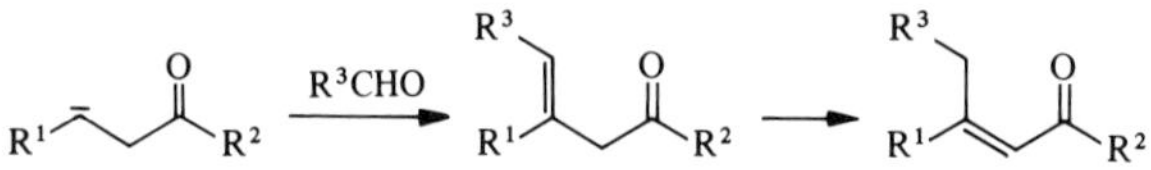

118 *119* *120*

The homologue of the specific enolate is the homoenolate *118*. This synthon obviously implies regiospecificity, but it is more important that umpolung of reactivity[9] is required as well. Enolates condense with carbonyl compounds to give conjugated enones and so the first application of homoenolates would be a similar condensation to give β,γ-unsaturated carbonyl compounds *119*. However, these easily give conjugated enones *120* and so homoenolates can also be used in an unconventional enone synthesis. An example of a homoenolate equivalent is Corey's cyclopropyl anion *121* used in the synthesis[74] of the β,γ-unsaturated aldehyde *122*.

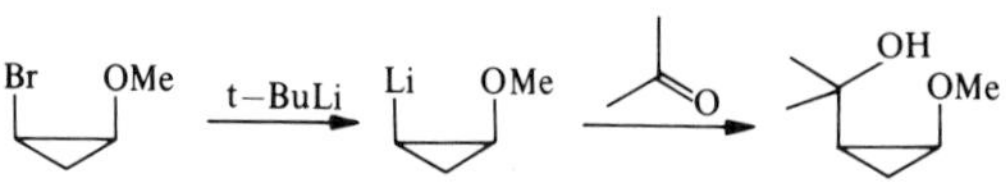

121

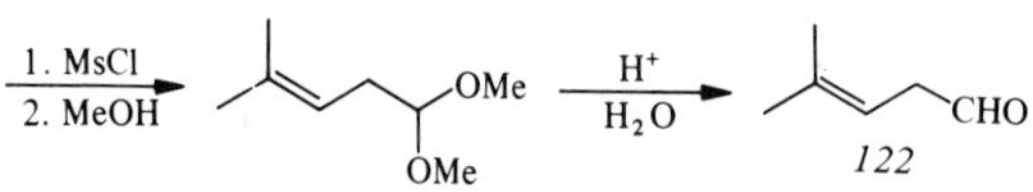

122

The simplest homoenolate equivalents are perhaps the anions *123* of allyl sulphides[17, 33, 54)], ethers[47)], or amines[75)], when they can be persuaded to react in the γ-position with electrophiles, as the heteroatom X in *123* becomes vinyl X in the product *124* and hence, after hydrolysis, a carbonyl group.

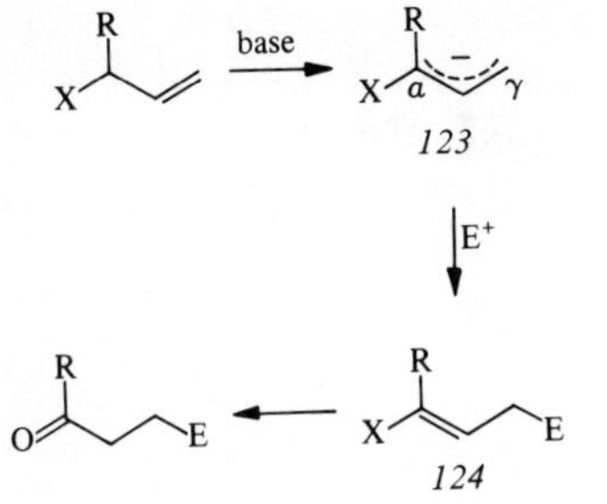

Alternatives include protected carbonyl compounds with anion-stabilising groups in the β-position such as the Grignard reagent[76)] *125* or the sulphone[77)] *126* used in prostaglandin syntheses. We have developed[78, 79)] general routes to make virtually any substituted versions of the corresponding phosphine oxide *127*.

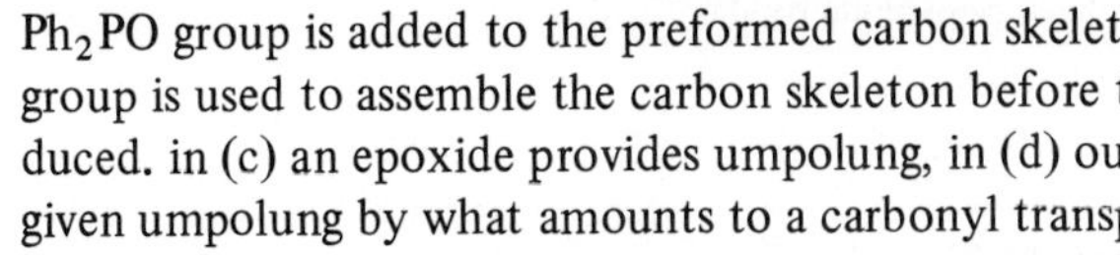

These routes are summarised in Scheme 3. In the first two, (a) and (b), the Ph_2PO group is added to the preformed carbon skeleton. In the rest, the Ph_2PO group is used to assemble the carbon skeleton before the carbonyl group is introduced. in (c) an epoxide provides umpolung, in (d) our allyl phosphine oxides *15* are given umpolung by what amounts to a carbonyl transposition: [1,2] or [1,3] depending on whether *26* or *27* is the starting material. The carbonyl transposition in (e), is more straightforward.

These routes are not equally good: (a) is difficult unless R^2 = H, (b), (c), and (e) are generally reliable, as is (d) *via* hydroboration. The epoxide rearrangement in (d) is capricious and has worked for us in high yield only for $R^1 = R^2$ = Me.

The reagents *128–130* are stable crystalline compounds which form anions easily with BuLi. Addition to aldehydes or ketones give protected β,γ-unsaturated carbonyl compounds, e. g. *131,* after completion of the Horner-Wittig reaction, and hence the ketones *132* themselves. These methods have not yet been used in natural product synthesis.

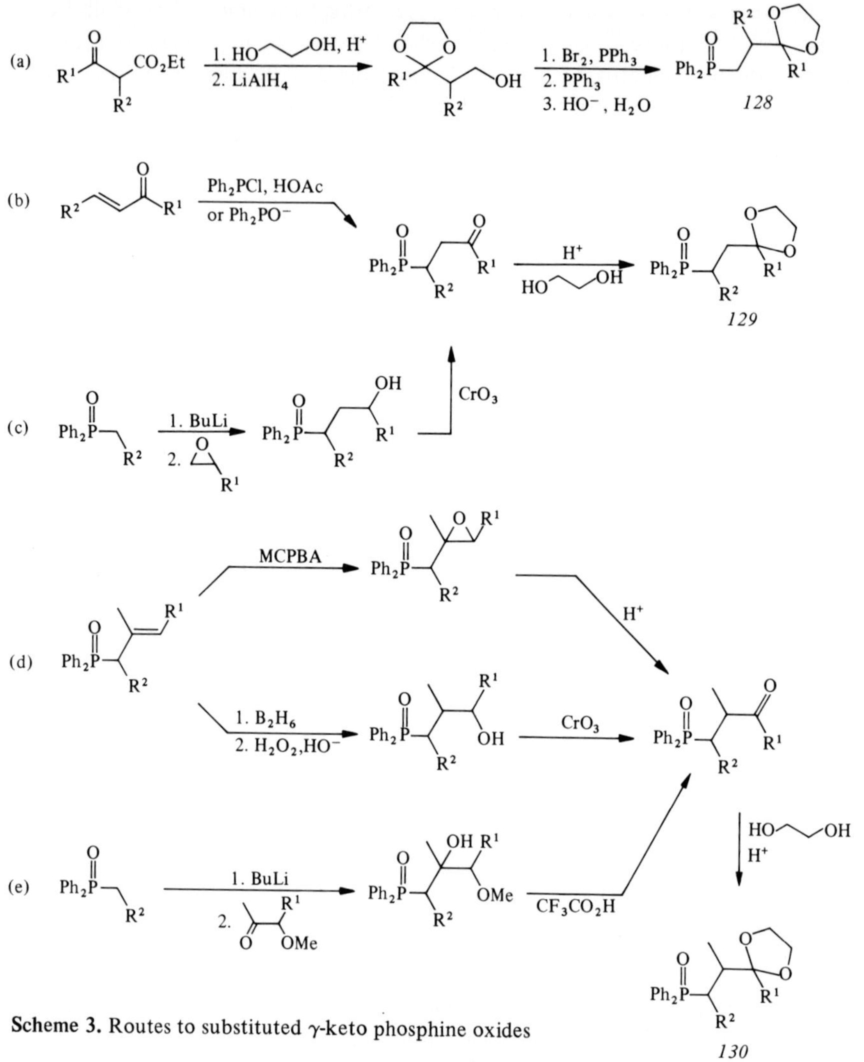

Scheme 3. Routes to substituted γ-keto phosphine oxides

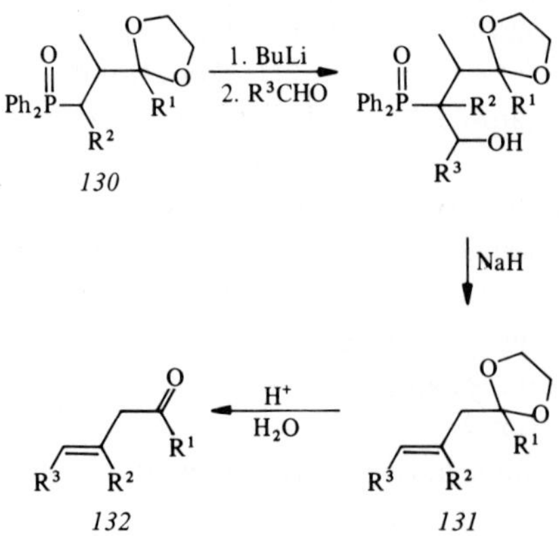

7 Reagents for the Synthons $^{-}$C=C–CO

The unsaturated version of the homoenolate synthon, the β-acyl vinyl anion[80] *133*, can be generated simply in the carboxylic acid series from the halide *134*, and has been used in butenolide *135* synthesis[81].

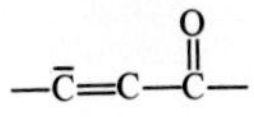

133

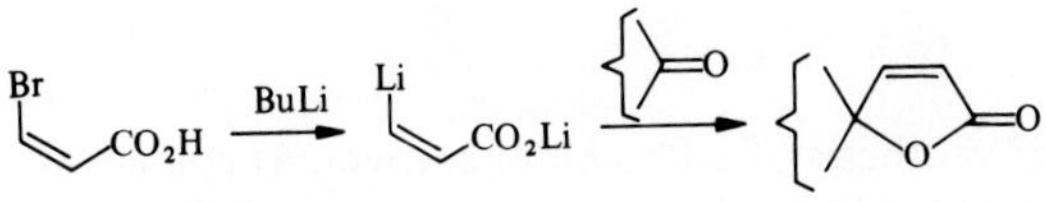

134 *135*

A more general solution to the problem is to use an allyl anion substituted on both ends with a heteroatom *136*. The anion *137* may then react regiospecifically with electrophiles to give products *138* with a masked carbonyl group (vinyl Y in *138*) and a leaving group (X in *138*) so that hydrolysis gives the enone *139*.

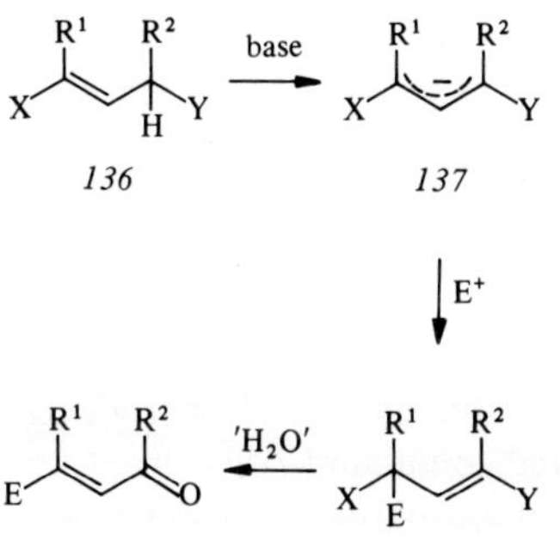

136 *137* *139* *138*

To be successful, this approach requires a general synthesis of *136* and regiospecific reaction of *137* with electrophiles. If *137* is symmetrical ($R^1 = R^2$, X = Y), the latter question is avoided, and Corey's reagent[82] *140* corresponding to the synthon *141* is based on this principle. The reagent *140* has been combined with epoxides to make[83] $PGF_{2\alpha}$ and 9(0) methano prostacyclin[84].

140 *141*

A more general route is that of Cohen[80], in which the enone itself is used as the starting material, and the reagents, e. g., *142*, shown to take part in Michael reactions.

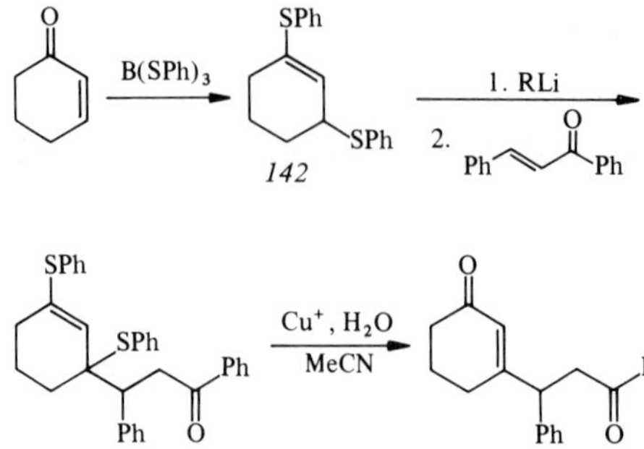

142

Since we had available a general synthesis of α-PhS ketones (Sect. 4) and a way of converting ketones into vinyl sulphides (Sect. 3.2.1), a combination of these two offered a general synthesis of the reagents *143*. The reaction was reasonably successful[49)], but a great deal of starting material was recovered, probably because the acidic proton in the α-PhS ketone is attacked by the anion. In addition, Cohen[80)] has now shown that anions of unsymmetrical reagents *143* ($R^1 = R^2$) show disappointing selectivity in their reactions with electrophiles.

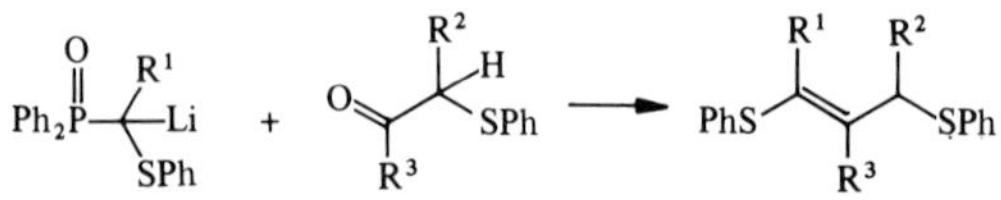

143

One way out of both these difficulties is to replace one PhS group in *143* with OMe and use reagents *144* which are intrinsically unsymmetrical. The α-methoxyl carbonyl compound no longer has a markedly acidic proton and the Horner-Wittig reaction[49)] gives good yields of *144*. The unsubstituted compound *144*, $R^1 = R^2 = H$, has been used in alkylations to give α,β-unsaturated aldehydes in good yield[85)]. Reaction occurs entirely α to PhS and γ to MeO leaving an easily hydrolysed vinyl ether as the product. The substituted reagents *144* must be carefully purified as PhSH easily displaces MeOH to give the less useful bis-(phenylthio) compounds.

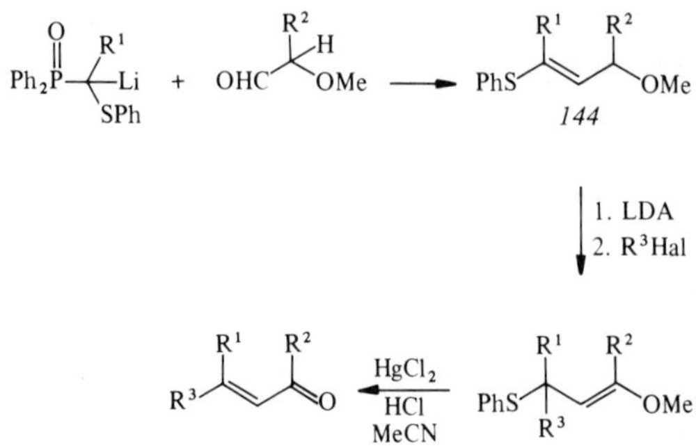

8 Phenylthio Butadienes for the Diels-Alder Reaction

Dienes substituted with heteroatoms, X or Y in *145,* give allyl (Y in *146)* or vinyl (X in *146*) derivatives by the Diels-Alder reaction. The heteroatom(s) not only provide latent functionality in the product but also control the regiochemistry of the Diels-Alder reaction itself[86]. In Danishefsky's vernolepin synthesis[87], a compound of type *146* was converted into an enone *147* by hydrolysis since *146* is a 1,3-disubstituted allyl compounds like the enone precursors in the last section.

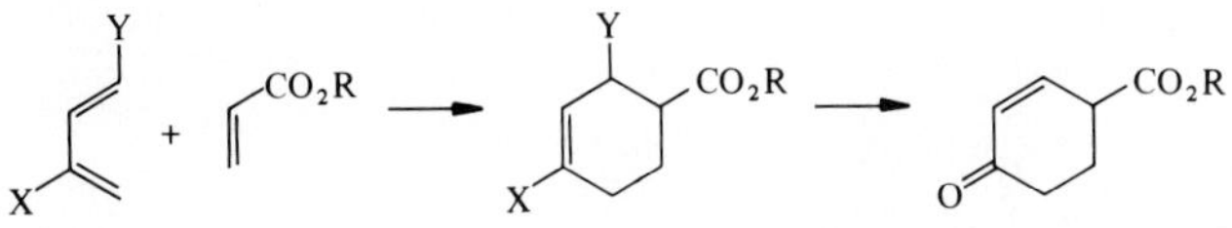

145 *146* *147*

Substituents X and Y based on sulphur are particularly interesting. We have already seen that vinyl sulphides are latent ketones, that allyl sulphides form useful anions and that their sulphoxides give allyl alcohols. Evans[17] has used cyclic allyl sulphoxide anions in an allyl alcohol synthesis, and has used the sulphoxide *148* in a Diels-Alder synthesis[88] of the hasubanan derivative *149.*

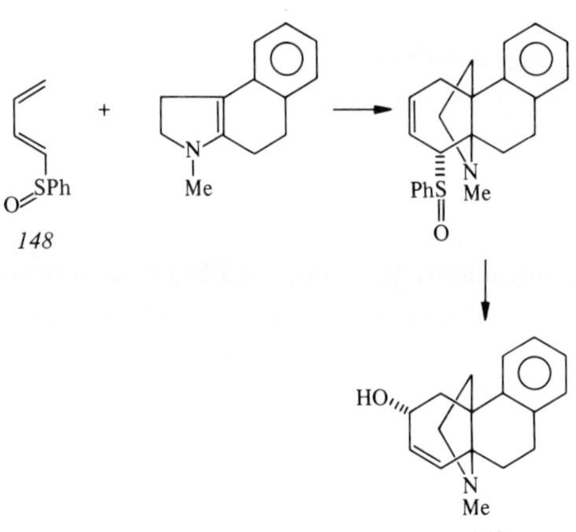

148 *149*

This synthesis emphasises another advantage of sulphur-based substituents in the Diels-Alder reaction. The PhS group is electron-donating – an $\dot{\text{X}}$ group[86] – and is more powerful than RO in determining the orientation of cycloadditions[89]. Oxidation to PhSO, as in *148,* makes it electron withdrawing – a Z substituent[86] – so that it will react with electron-rich dienophiles in a predictable way, as with the enamine to make *149.*

There are few syntheses of either 1- or 2-PhS butadienes, those of Cohen[90] being the most general. There are however, two general approaches to 1-PhS butadienes *150* based on the Horner-Wittig reaction. Either our[49] PhS-substituted phosphine oxides *79* might react with enals *151* or γ-PhS substituted allyl phosphine oxides *152* might react with aldehydes.

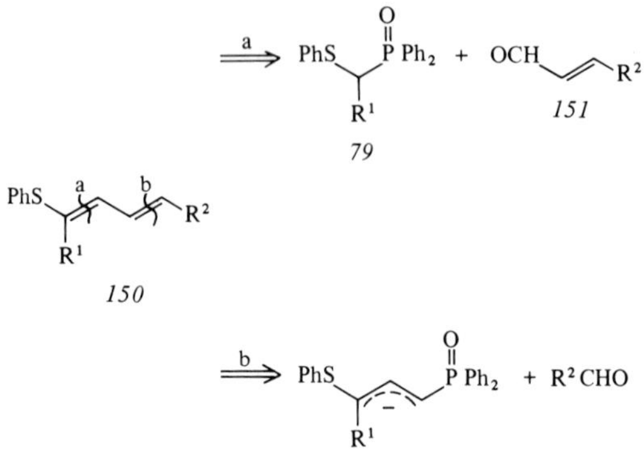

79 *151* *150* *152*

In the event, both methods are successful[31]. Reagents *79* add only 1,2 to enals, possibly because the Horner-Wittig reaction goes to completion under the reaction conditions. Dienes *153* (92% yield) and *154* (96%) can be made by this method[31, 49], and diene *154* gives the Diels-Alder adduct *155* in 90% yield.

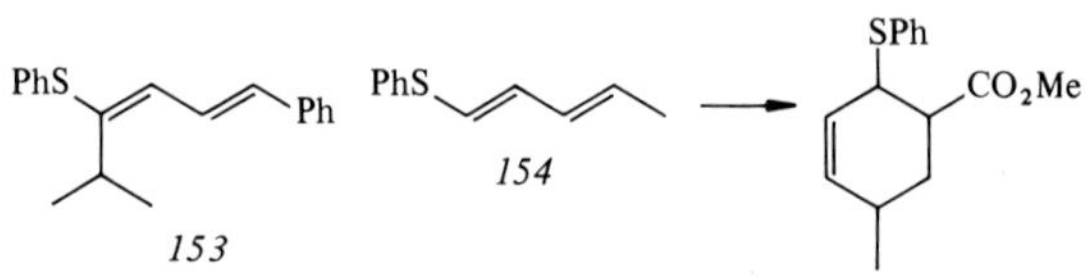

153 *154* *155*

Sulphenylation of allyl phosphine oxide anions gives only γ-addition, in contrast to the predominantly α attack with other electrophiles (Sect. 2.1). This is because the product of α attack can isomerise by a [1,3] shift[22]. The product of the reaction is a mixture of allyl isomers *156* and *157*. These both give the anion *158* on treat-

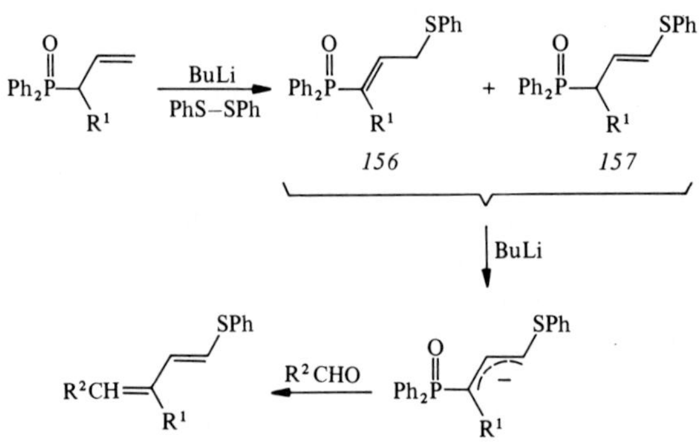

156 *157* *159* *158*

ment with BuLi and hence the dienes *159* by the Horner-Wittig reaction. The anion *158* then reacts with carbonyl compounds α to Ph_2PO and γ to PhS, following the usual trends for such substituents[30, 33]. Dienes made by this route[31], e. g. *159,* R^1 = Me, R^2 = Ph, 68% yield, have a different substitution pattern from that of *153* or *154.*

We made[31] 2-PhS butadienes *162* by thermal elimination on the sulphoxides *161* (cf. *110),* the rearrangement products which we had previously used to make 2-PhS allyl alcohols. Hence, *160,* R^1 = Ph, R^2 = Me, gave the corresponding diene *162* in 62% overall yield, and the diene *163* gave the expected Diels-Alder adduct *164* in 78% yield.

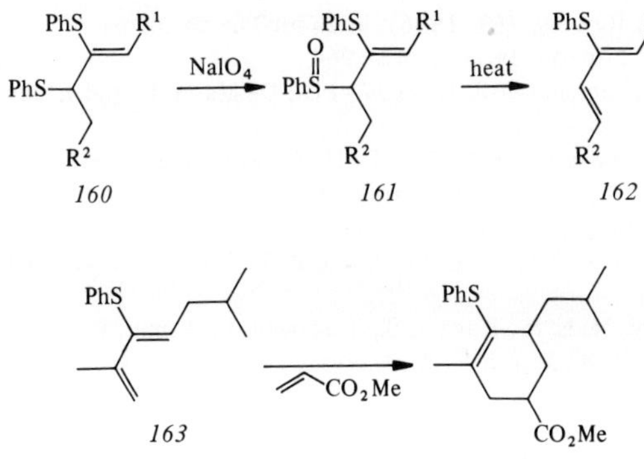

160 *161* *162*

163 *164*

Acknowledgements. It is a pleasure for me to thank my co-workers, particularly A. H. Davidson, P. Brownbridge, P. Blatcher, C. Earnshaw, J. I. Grayson, and R. S. Torr for carrying out the work described here.

9 References

1. Corey, E. J.: Quart. Rev. *25,* 455 (1971)
2. Corey, E. J., Cramer, R. D., Howe, W. J.: J. Am. Chem. Soc. *94,* 440 (1972); Corey, E. J., Jorgensen, W. L.: J. Am. Chem. Soc. *98,* 189, 203 (1976); Corey, E. J., Orf, H. W., Pensak, D. A.: J. Am. Chem. Soc. *98,* 210 (1976) and references therein
3. Warren, S.: Designing organic syntheses. Chichester: Wiley 1978
4. Corey, E. J., Balanson, R. D.: J. Am. Chem. Soc. *96,* 6516 (1974)
5. Oppolzer, W., Godel, T.: J. Am. Chem. Soc. *100,* 2583 (1978)
6. Corey, E. J., Ohno, M., Mitra, R. B., Vatakencherry, P.: J. Am. Chem. Soc. *86,* 478 (1964); Corey, E. J., Howe, W. J., Orf, H. W., Pensak, D. A., Peterson, G.: J. Am. Chem. Soc. *97,* 6116 (1975)
7. Corey, E. J.: Pure Appl. Chem. *14,* 19 (1967)
8. Corey, E. J.: Chemical Society Symposium on Synthesis in Organic Chemistry, Oxford 1977
9. Seebach, D.: Angew. Chem., Int. Ed. *18,* 239 (1979)
10. Wittig, G.: Top. Curr. Chem. *67,* 2 (1976); D'Angelo, J.: Tetrahedron *32,* 2979 (1976)
11. Stork, G.: Pure Appl. Chem. *43,* 553 (1975)
12. Rasmussen, J. K.: Synthesis *1977,* 91

13. Seebach, D., Kolb, M.: Chem. and Ind. *1974,* 687
14. Lever, O. W.: Tetrahedron *32* 1943 (1976)
15. Seebach, D., Corey, E. J.: J. Org. Chem. *40,* 231 (1975)
16. See, for exapmle, Vedejs, E., Fuchs, P. L.: J. Org. Chem., *36,* 366 (1971); Torii, S., Uneyama, K., Ishihara, M.: J. Org. Chem., *39,* 3645 (1974)
17. Evans, D. A., Andrews, G. C.: Acc. Chem. Res. *7,* 147 (1974)
18. Trost, B. M., Hiroi, K., Kurozumi, S.: J. Am. Chem. Soc. *97,* 438 (1975) and references therein
19. Watt, D. S., Corey, E. J.: Tetrahedron Lett. *1972,* 4651
20. Oppolzer, W., Mahalanabis, K. K.: Tetrahedron Lett. *1975,* 3411
21. Warren, S.: Acc. Chem. Res. *11,* 401 (1978)
22. Brownbridge, P., Warren, S.: J. Chem. Soc., Perkin I, *1976,* 2125. For a different interpretation of the mechanism of this reaction, see Kwart, H., George, T. J.: J. Am. Chem. Soc. *99,* 5214 (1977)
23. de Wolfe, R. H., Young, W. G.: Chem. Rev. *56,* 769 (1956): idem in: The chemistry of alkenes, pp. 688–731. Patai, S., (ed.). London: Interscience 1964
24. De la Mare, P. B. D.: Molecular rearrangements, Vol. 1, pp 27–110. De Mayo, P., (ed.). New York: Interscience 1963
25. Gompper, R., Wagner, H.-U.: Angew. Chem. Int. Ed. *15,* 321 (1976). For a recent leading reference, see Linstrumelle, G., Lorne, R., Dang, H. P.: Tetrahedron Lett. *1978,* 4069
26. Davidson, A. H., Warren, S.: J. Chem. Soc. Perkin 1, *1976,* 639
27. Davidson, A. H., Fleming, I., Grayson, J. I., Pearce, A., Snowden, R. L., Warren, S.: J. Chem. Soc., Perkin 1, *1977,* 550
28. Lythgoe, B., Moran, T. A., Nambudiry, M. E. N., Ruston, S., Tideswell, J., Wright, P. W.: Tetrahedron Lett. *1975,* 3863; Lythgoe, B., Moran, T. A., Nambudiry, M. E. N., Ruston, S.: J. Chem. Soc., Perkin I, *1976,* 2386
29. Clough, J. M., Pattenden, G.: Tetrahedron Lett. *1978,* 4159
30 Davidson, A. H., Earnshaw, C., Grayson, J. L., Warren, S.: J. Chem. Soc., Perkin 1, *1977,* 1452
31. Blatcher, P., Grayson, J. I., Warren, S.: J. Chem. Soc. Chem. Commun. *1978,* 657
32. Arndt, R. R., Warren, S.: Tetrahedron Lett. *1978,* 4089
33. Brownbridge, P., Warren, S.: J. Chem. Soc., Perkin 1, *1977,* 1131; 2272
34. Brownbridge, P., Fleming, I., Pearce, A., Warren, S.: J. Chem. Soc., Chem. Commun. *1976,* 751
35. Evans, D. A., Crawford, T. C., Fujimoto, T. T., Thomas, R. C.: J. Org. Chem. *39,* 3176 (1974)
36. Davidson, A. H., Earnshaw, C., Torr, R., Warren, S.: unpublished
37. Bayer, O.: Methoden der Organischen Chemie (Houben-Weyl), B and VII, Teil 1, pp 326–329. Sauerstoff Verbindungen 11, Aldehyde, Stuttgart: Thieme 1954
38. Kornfeld, E. C., Fornefeld, E. J., Kline, G. B., M nn, M. J., Morrison, D. E., Jones, R. G., Woodward, R. B.: J. Amer. Chem. Soc. *78,* 308 (1956)
39. Wittig, G., Schlosser, M.: Chem. Ber. *94,* 1373 (1961); Wittig, G., Böll, W., Krück, K-H.: Chem. Ber. *95,* 2514 (1962)
40. Pettit, G. R., Green, B., Dunn, G. L., Sunder-Plassmann, P.: J. Org. Chem. *35,* 1385 (1970)
41. Earnshaw, C., Wallis, C. J., Warren, S.: J. Chem. Soc., Chem. Commun. *1977,* 314; Perkin 1, *1979,* 3099
42. Schlosser, M., Tuong, H. B.: Chimia *30,* 197 (1976)
43. Harley-Mason, J., Wallis, C. J.: unpublished
44. Hudrlik, P. F., Hudrlik, A. M., Rona, R. J., Misra, R. N., Withers, G. P.: J. Am. Chem. Soc. *99,* 1993 (1977)
45. Muthukrishnan, R., Schlosser, M.: Helv. Chim. Acta *59,* 13 (1976); Hartman, J., Stähle, M., Schlosser, M.: Synthesis *1974,* 888
46. Baldwin, J. E., Höfle, G. A., Lever, O. W.: J. Am. Chem. Soc. *96,* 7125 (1974)
47. Evans, D. A., Andrews, G. C., Buckwalter, B.: J. Am. Chem. Soc. *96,* 5560 (1974); Still, W. C., Macdonald, T. L.: J. Am. Chem. Soc. *96,* 5561; J. Org. Chem.: *41,* 3620 (1976)
48. Blatcher, P., Grayson, J. I., Warren, S.: J. Chem. Soc., Chem. Commun. *1976,* 547

49. Grayson, J. I., Warren, S.: J. Chem. Soc., Perkin 1, *1977,* 2263; Earnshaw, C., Grayson, J. I., Warren, S.: J. Chem. Soc., Perkin 1, *1979,* 1506
50. Cookson, R. C., Parsons, P. J.: J. Chem. Soc., Chem. Commun. *1976,* 990; *1978,* 821
51. Mukaiyama, T., Fukuyama, S., Kumamoto, T.: Tetrahedron Lett. *1968,* 3787
52. Corey, E. J., Shulman, J. I.: J. Org. Chem. *35,* 777, (1970)
53. Mc Guire, H. M., Odom, H. C., Pinder, A. R.: J. Chem. Soc. Perkin 1, *1974,* 1879
54. Grobel, B.-T., Seebach, D.: Synthesis *1977,* 357
55. Seebach, D.: Synthesis *1969,* 17
56. Smith, R. G., Daterman, G. E., Daves, G. D.: Science *188,* 63 (1975)
57. Blatcher, P., Warren, S.: J. Chem. Soc. Perkin 1, *1979,* 1074
58. Gerber, U., Widmer, U., Schmid, R., Schmid, H.: Helv. Chim. Acta *61,* 83 (1978)
59. Monteiro, H. J.: J. Org. Chem. *42,* 2324 (1977)
60. Kieczykowski, G. R., Pognowski, C. S., Richman, J. E., Schlessinger, R. H.: J. Org. Chem. *42,* 175 (1977)
61. Seebach, D., Teschner, M.: Chem. Ber. *109,* 1601 (1976)
62. Trost, B. M., Saltzman, T. N., Hiroi, K.: J. Am. Chem. Soc. *98,* 4887 (1976)
63. Murai, S., Kuroki, Y., Hasegawa, K., Tsutsumi, S.: J. Chem. Soc., Chem. Commun. *1972,* 946
64. Blatcher, P., Warren, S.: J. Chem. Soc., Chem. Commun. *1976,* 1055
65. Kurozumi, S., Toro, T., Kobayashi, M., Ishimoto, S.: Syn. Commun. *1977,* 427
66. Grieco, P. A.: Synthesis *1975,* 67
67. Brownbridge, P., Warren, S.: J. Chem. Soc., Chem. Commun. *1977,* 465
68. Rao, Y. S.: Chem. Revs. *76,* 625 (1976)
69. Trost, B. M., Hiroi, K., Holy, M.: J. Am. Chem. Soc. *97,* 5873 (1975)
70. Kano, S., Yokomatsu, T., Shibuya, S.: J. Chem. Soc., Chem. Commun. *1978,* 785
71. Grieco, P. A., Hiroi, K.: Tetrahedron Lett. *1973,* 1831
72. Trost, B. M., Hiroi, K.: J. Am. Chem., Soc. *97,* 6911 (1975); Trost, B. M., Massiot, G. S.: J. Am. Chem., Soc. *99,* 4405, (1977)
73. Blatcher, P., Warren, S.: Tetrahedron Lett. *1979,* 1247
74. Corey, E. J., Ulrich, P.: Tetrahedron Lett. *1975,* 3685
75. Martin, S. F., Du Priest, M. T.: Tetrahedron Lett. *1976,* 3105
76. Ponaras, A. A.: Tetrahedron Lett. *1976,* 3105
77. Kondo, K., Tunemoto, D.: Tetrahedron Lett. *1975,* 1397
78. Bell, A., Davidson, A. H., Earnshaw, C., Norrish, N. K., Torr, R. S., Warren, S.: J. Chem. Soc. Chem. Commun. *1978,* 988
79. Torr, R. S., Trowbridge, D. B., Warren, S.: unpublished
80. Cohen, T., Bennett, D. A., Mura, A. J.: J. Org. Chem. *41,* 2506 (1976)
81. Caine, D., Frobese, A. S.: Tetrahedron Lett. *1978,* 5167
82. Corey, E. J., Erickson, B. W., Noyori, R.: J. Am. Chem. Soc. *93,* 1724 (1971)
83. Corey, E. J., Noyori, R.: Tetrahedron Lett. *1970,* 311
84. Shibasaki, M., Ueda, J., Ikegami, S.: Tetrahedron Lett. *1979,* 433
85. Wada, M., Nakamura, H., Taguchi, T., Takei, T.: Chemistry Lett. *1977,* 345
86. Fleming, I.: Frontier orbitals and organic chemical reactions. p. 86. London: Wiley 1976
87. Danishefsky, S., Kitahara, T., Schuda, P. F., Etheridge, S. J.: J. Am. Chem. Soc. *98,* 3028 (1976)
88. Evans, D. A., Bryan, C. A., Sims, C. L.: J. Am. Chem. Soc. *94,* 2891 (1972)
89. Trost, B. M., Ippen, J., Vladuchick, W. C.: J. Am. Chem. Soc. *99,* 8116 (1977)
90. Cohen, T., Mura, A. J., Shull, D. W., Fogel, E. R., Ruffner, R. J., Falck, J. R.: J. Org. Chem. *41,* 3218 (1976)

Received July 19, 1979

Applications of Palladium-Catalyzed or Promoted Reactions to Natural Product Syntheses

Jiro Tsuji

Tokyo Institute of Technology, Meguro, Tokyo 152, Japan

Table of Contents

1 Introduction

One of the remarkable advances in organic chemistry made in the last decade is the discovery of numerous organic reactions involving transition metal complexes. Transformations which are not possible by conventional methods of organic chemistry can be achieved sometimes by using transition metal complexes either as catalysts or stoichiometric reagents. A number of elegant and short-step syntheses of some natural products have been accomplished by applying transition-metal promoted or catalyzed reactions. In this sense, palladium is not an exception.

In the last 20 years, after the invention of the *Wacker* process to produce acetaldehyde from ethylene using $PdCl_2/CuCl_2$ as a catalyst[1)], many organic reactions either promoted or catalyzed by palladium compounds have been discovered[2)]. Some of them are potentially useful for organic synthesis, and a number of ingenious applications of these reactions to natural product syntheses have been reported. In this review, natural product syntheses using palladium compounds either as catalysts or reagents in key steps are summarized, in the hope of stimulating further development in this field.

Organic reactions involving palladium compounds can be classified into two main types. The first type involves stoichiometric oxidative reactions with $Pd^{2\oplus}$ compounds, $Pd^{2\oplus}$ being reduced to Pd^0. However, in some cases as in the *Wacker* process, the reduced Pd^0 is reoxidized to $Pd^{2\oplus}$ by appropriate oxidizing agents, such as $CuCl_2$ and benzoquinone, thus enabling the use of $Pd^{2\oplus}$ in catalytic amounts. However, reoxidation of Pd^0 may not always be readily conducted; thus, in many cases, stoichiometric amounts of $Pd^{2\oplus}$ are consumed. The stoichiometric consumption of rather expensive $Pd^{2\oplus}$ compounds being intolerable in preparative organic chemistry represents a serious limitation of reactions involving these compounds. Typical Pd^{2+} promoted reactions, surveyed here include:

Wacker-type oxidative reactions of olefins with nucleophiles, reactions of π-allylpalladium complexes with nucleophiles, reactions based on chelation, and transmetallation of organomercury compounds.

Reactions of the second type are carried out with palladium compounds or complexes of either bivalent or zero-valent states. Since these reactions proceed catalytically without using reoxidants they are more useful than the stoichiometric processes. Telomerization of conjugated dienes, reactions of allylic and alkenyl esters and ethers, and various organic halides belong to this type.

2 Oxidative Reactions with $Pd^{2\oplus}$ Compounds

2.1 Reactions of Olefins with Nucleophiles

Nucleophilic substitution and addition reactions of olefins are possible with $Pd^{2\oplus}$ salts. A typical example is the formation of acetaldehyde by the reaction of ethylene with water (*Wacker* reaction). As nucleophiles, water, alcohols, phenols, carboxylic acids, amines, enamines, carbanions derived from active methylene compounds, and carbon monoxide react with olefins with stoichiometric consumption of $Pd^{2\oplus}$ salts.

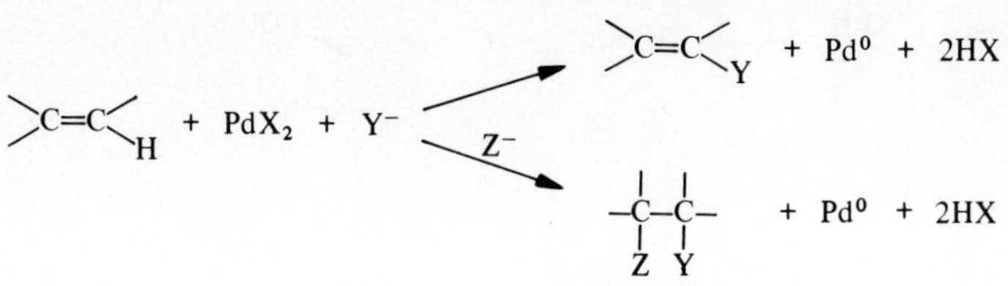

Some reaction products from ethylene and nucleophiles are shown in Scheme 1. In some cases, the reaction can be made catalytic by using appropriate reoxidants.

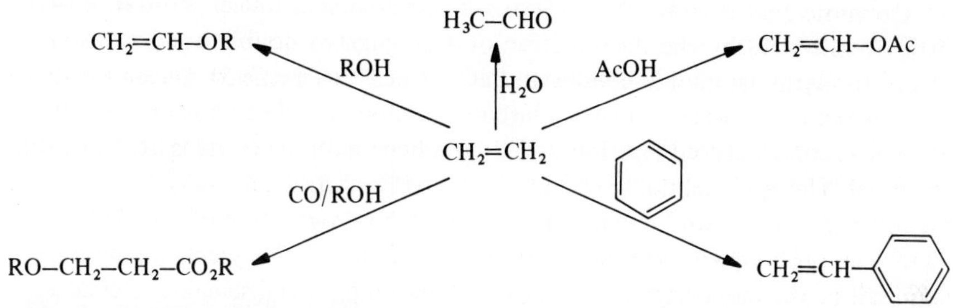

Scheme 1

Higher olefins are oxidized selectively with $PdCl_2$ in aqueous organic solvents to methyl ketones without forming aldehydes. The reaction is carried out in DMF using a catalytic amount of $PdCl_2$ together with $CuCl_2$ or CuCl and benzoquinone as reoxidants[3, 4]. This is a useful reaction, and terminal double bonds can be regarded as masked ketones.

$$RCH{=}CH + 1/2\ O_2 \xrightarrow[CuCl_2]{PdCl_2} R{-}\overset{\overset{\large O}{\|}}{C}{-}CH_3$$

The terminal double bond of 9-decenoic acid (*1*), prepared from commercially available 10-undecenoic acid, is oxidized with $PdCl_2/CuCl_2$ to give 9-oxodecanoic acid (*2*), which is converted to 3-(6-methoxycarbonylhexyl) cyclopentane-1,2,4-trione (*3*). This is an important intermediate in the prostaglandin synthesis[5] (Scheme 2).

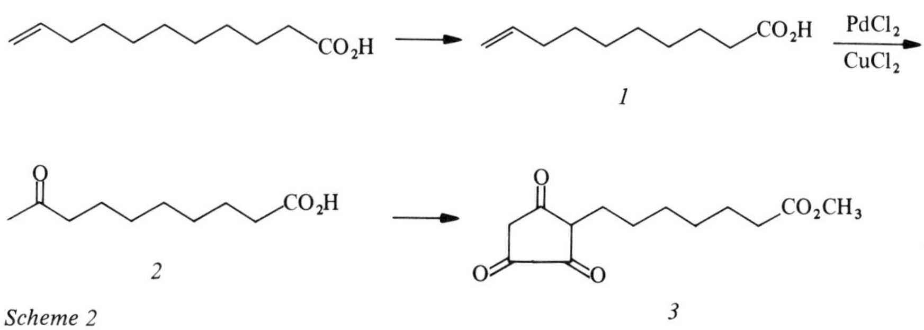

Scheme 2

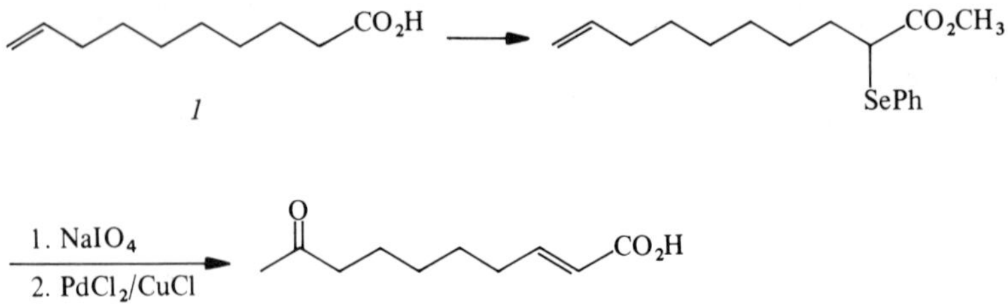

4

Scheme 3

9-Decenoic acid is converted to 9-oxo-2-decenoic acid [queen substance (*4*)] in 70% overall yield by the introduction of a conjugated double bond and oxidation of the terminal double bond with $PdCl_2/CuCl$[6] (Scheme 3). Queen substance (*4*) is prepared more easily from a 1,3-butadiene telomer[7]. The telomer *5* obtained by the palladium-catalyzed reaction of 1,3-butadiene with malonate is used as a starting material. The terminal double bond of *5* is oxidized to the δ, ϵ-unsaturated ketone *6*, the internal double bond remaining unaffected. Hydrogenation of the olefinic bond of *6* and subsequent treatment with potassium hydroxide yields the monopotassium salt of the monoester which is treated with diphenyl diselenide. Oxidative removal of the phenylselenenyl group affords queen substance (*4*) (Scheme 4).

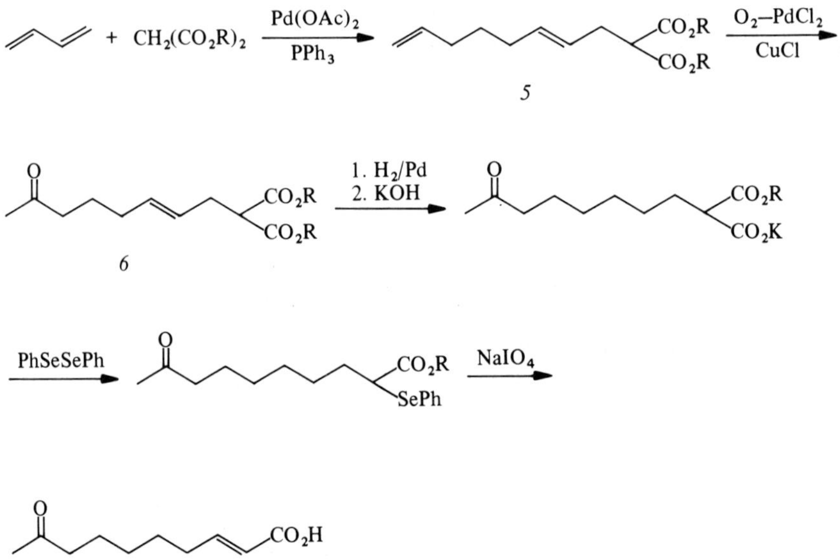

4

Scheme 4

The oxidation of the terminal double bond catalyzed by $PdCl_2/CuCl$ has been utilized in a general synthetis of 1,4- and 1,5-diketones which are important intermediates for five- and six-membered cyclic ketones[8]. At first, an allyl group is introduced into the α-position of ketones and then the terminal double bond is oxidized with

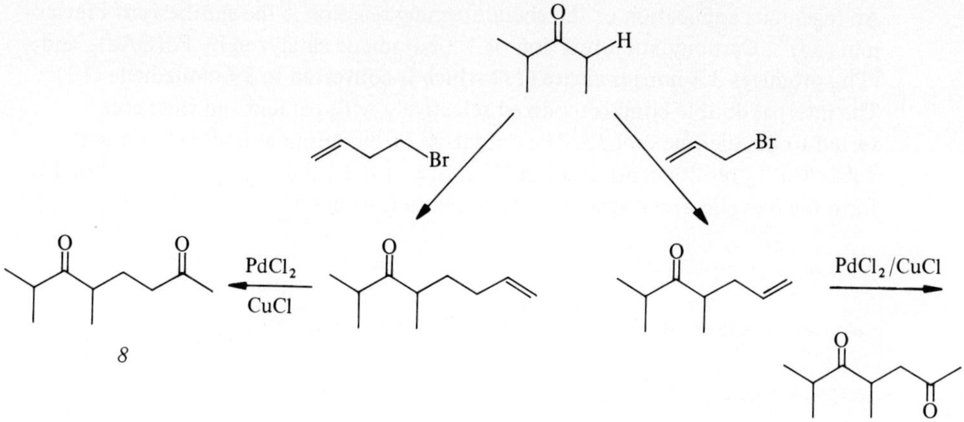

Scheme 5

$O_2/PdCl_2/CuCl$ to give 1,4-diketones *7*. Similarly, the introduction into ketones of the 1-butenyl group instead of the allyl group, followed by oxidation, affords 1,5-diketones *8* (Scheme 5). The described route to 1,4-diketones has been utilized in the dihydrojasmone synthesis[8]: The first step involves oxidation of 1-octene with $PdCl_2/CuCl$ to 2-octanone. After the introduction of the allyl group at the terminal position, the terminal double bond is unmasked to produce 2,5-undecanedione (*9*)

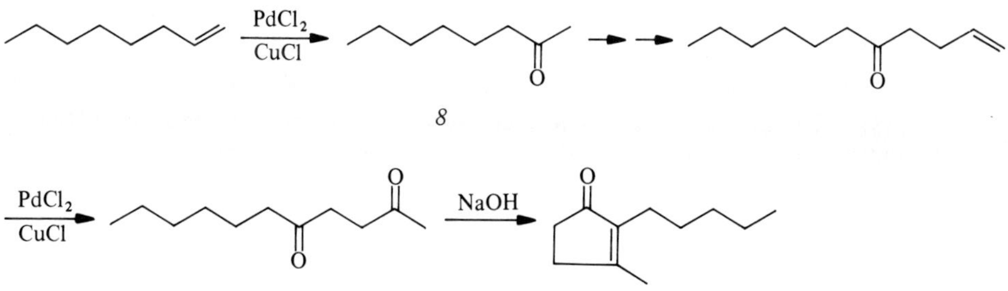

Scheme 6

which is cyclized to dihydrojasmone (*10*) (Scheme 6). For further examples see pp. 14–18, 23–25, 29. The oxidation of olefins with $PdCl_2$ in the presence of alcohols affords acetals as main products and vinyl ether as by-products

$$R{-}CH{=}CH_2 + ROH \xrightarrow{PdCl_2} R{-}\underset{OR}{\overset{OR}{\underset{|}{\overset{|}{C}}}}{-}CH_3 + R{-}\underset{OR}{\underset{|}{C}}{=}CH_2$$

An ingenious application of the acetal forming reaction is the synthesis of brevicomin (*14*)[9]. Carbonylation reaction of 1,3-butadiene catalyzed by $Pd(OAc)_2$ and PPh_3 produces 3,8-nonadienoate (*11*) which is converted to 1,6-nonadiene (12). The internal double bond is oxidized selectively with peracid and then converted to the olefinic diol *13*. The oxidation of the terminal double bond with $PdCl_2/CuCl_2$ results in intramolecular attack of the 1,2-diol at the double bond to form the bicyclic acetal system of brevicomin (*14*) in 45% yield (Scheme 7).

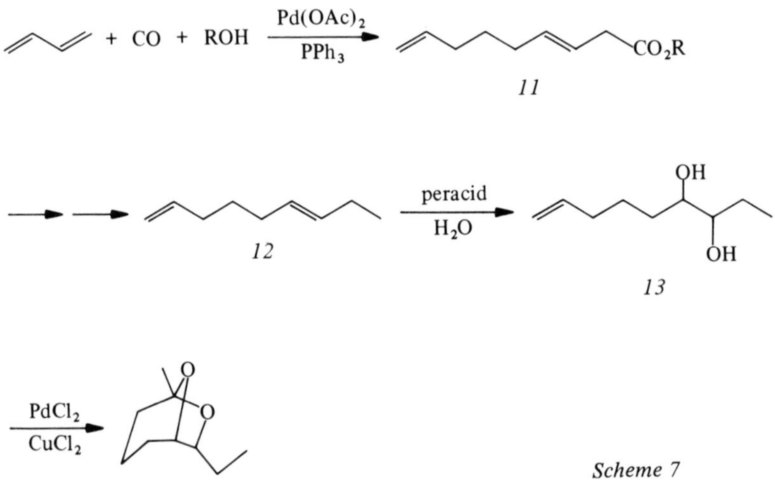

Scheme 7

A phenol group may reacts with double bonds to give phenyl substituted cyclic enol ethers. Thus, treatment of 2-(1-oxo-3-phenyl-2-propenyl) phenol (*15*) (2-hydroxychalcones) with $PdCl_2$ causes intramolecular attack of the hydroxy group at the olefinic double bond to form flavone (*16*)[10]. A different type of oxidation of phenols with $PdCl_2$

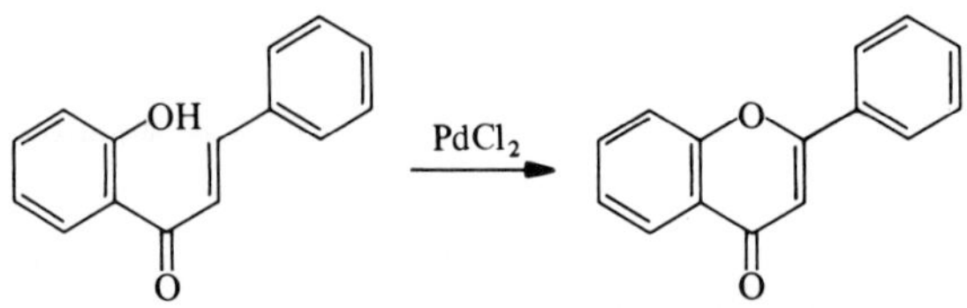

15 *16*

has been utilized in ingenious simple synthesis of carpanone[11]: Carpanone (*19*) is prepared in one step in 62% crude yield by oxidation of the 2-(1-propenyl)phenol derivative *17* with $PdCl_2$. This remarkable reaction is explained by the formation of an *o*-quinone derivative by one-electron oxidation with $PdCl_2$, followed by side chain radical coupling to give the intermediate *18*. Subsequent intramolecular cycloaddition yields carpanone (*19*). The nucleophilic amino group of substituted ureas

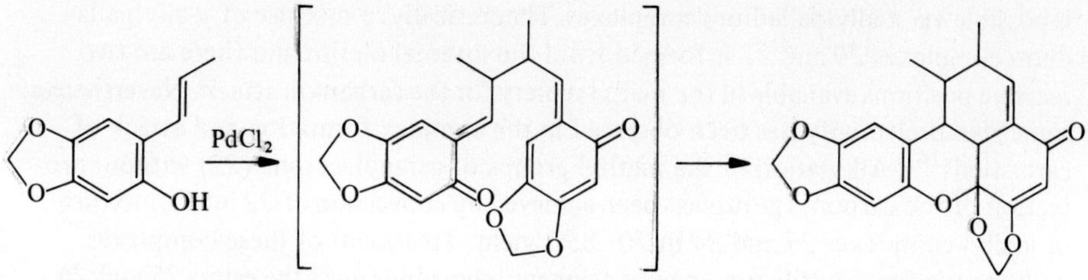

17 18 19

intramolecularly reacts with olefinic double bonds. Thus the reaction of acryloylurea promoted by $PdCl_2$ affords uracil[12].

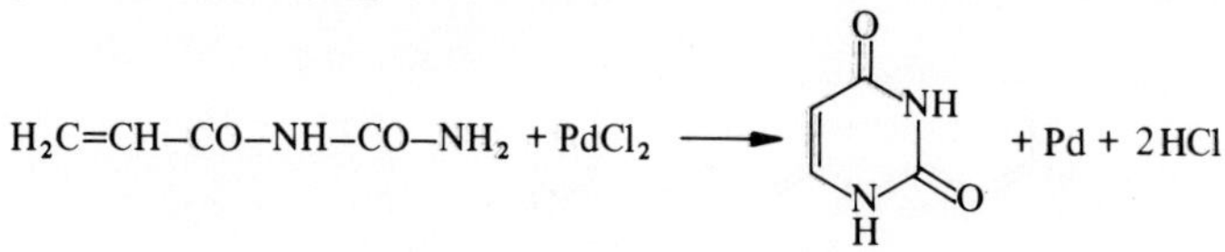

42%

2.2 Reaction of π-Allylpalladium Complexes

π-Allylpalladium complexes react with nucleophiles[13]. Especially, the reaction of carbanions, derived from active methylene compounds, with π-allylpalladium complexes offers a method of carbon-carbon bond formation. π-Allylpalladium com-

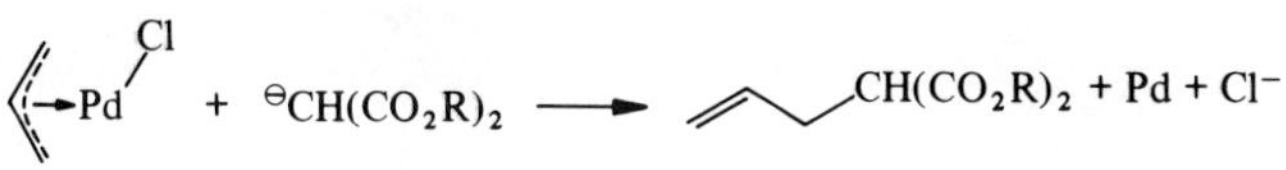

plexes can be prepared from allylic halides and alcohols. In addition, olefins can be converted to π-allylpalladium complexes by treatment with $PdCl_2$ in organic solvents like DMF. Thus, the alkylation of allylic positions of olefins with carbanions

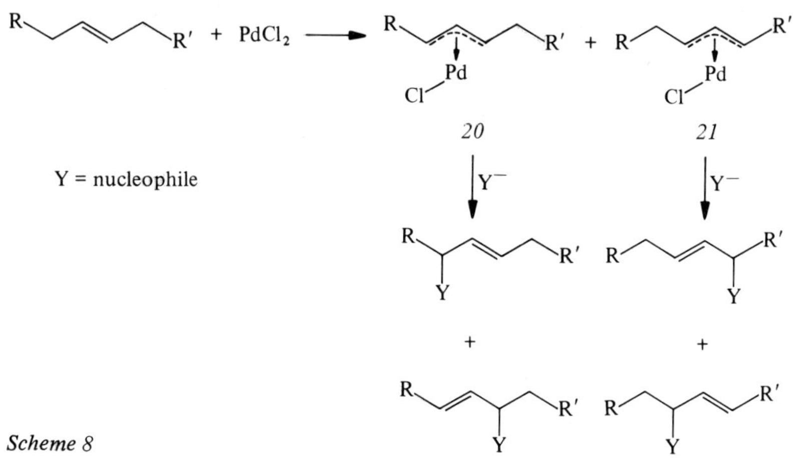

Scheme 8

is possible via π-allylpalladium complexes. Theoretically, a mixture of π-allylpalladium complexes *20* and *21* is formed from the internal olefins and there are two reactive positions available in the π-allyl moiety for the carbanion attack. Nevertheless, some stereoselectivity has been observed in the complex formation and attack of carbanions[14]. Alkylation of the methyl groups of geranylacetone (*22*) without protection of the carbonyl group has been achieved by conversion of *22* into a mixture of π-allyl complexes *23* and *24* in 70–85% yield. Treatment of these complexes with mesylacetate in the presence of triphenylphosphine gives the esters *25* and *26* in 24–85% yield. Finally, removal of the functional groups provides the methylated products *27*–*30*[15] (Scheme 9). Starting from methyl geraniate (*31*), pheromone of *Monarch* butterfly *33* was prepared via the π-allylpalladium complex *32* and sub-

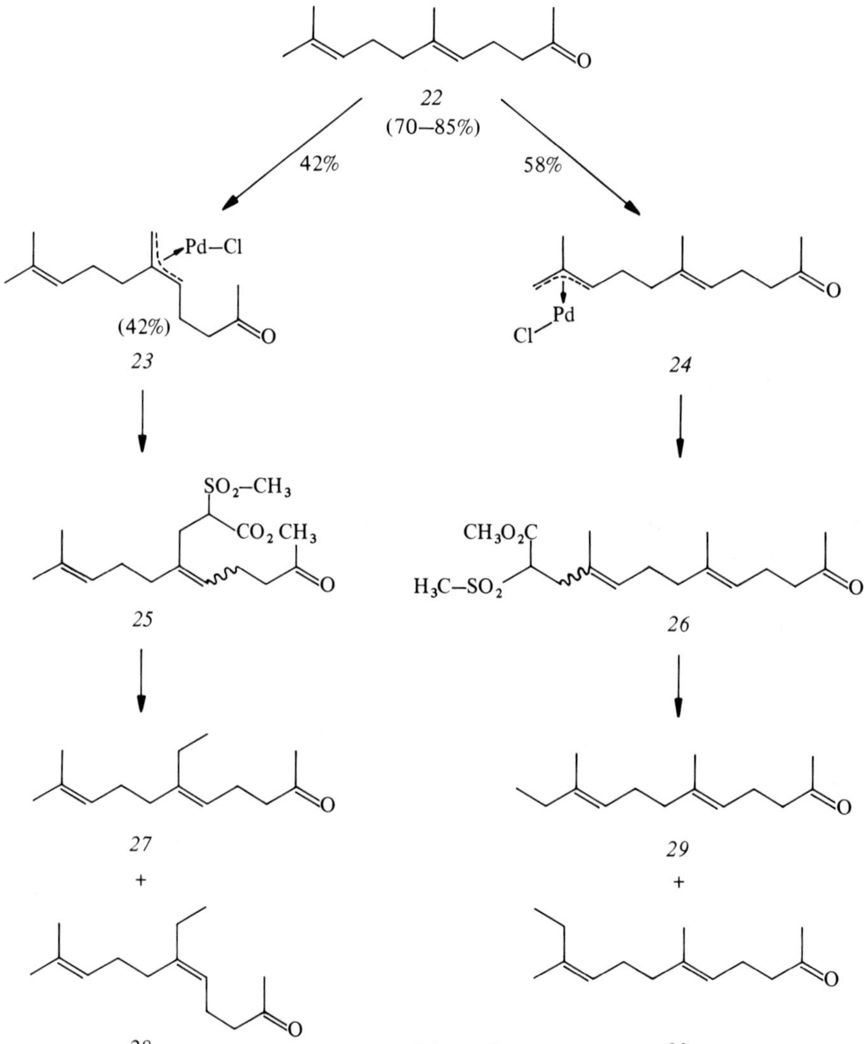

Scheme 9

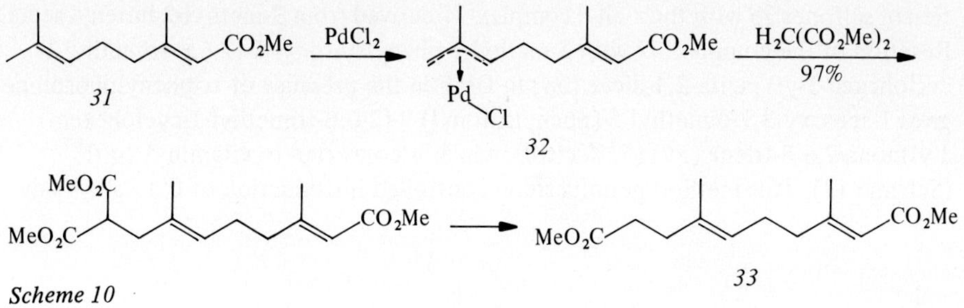

Scheme 10

sequent reaction with dimethyl malonate[16] (Scheme 10). Similarly, the 2-butenylation of methyl geraniate as described in Scheme 11 leads to farnesol (*34*)[16]. Another

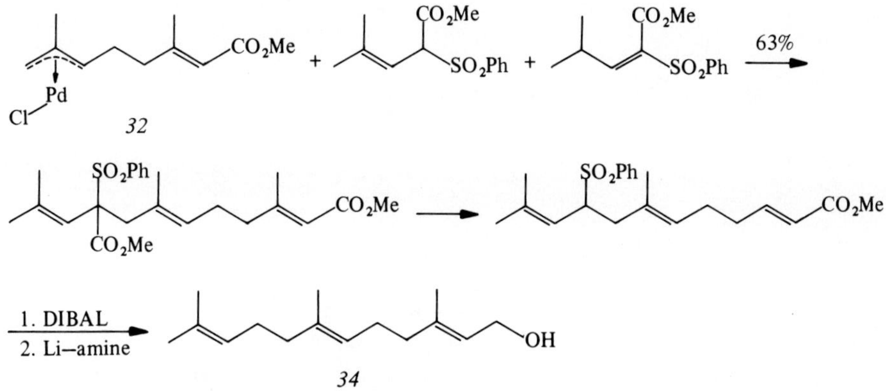

DIBAL = diisobutylaluminium hydride

Scheme 11

example is the formation of geranylgeraniol (*36*) from methyl farnesoate (*35*)[16] (Scheme 12). Vitamin A (*40*) and related compounds are synthesized by the reac-

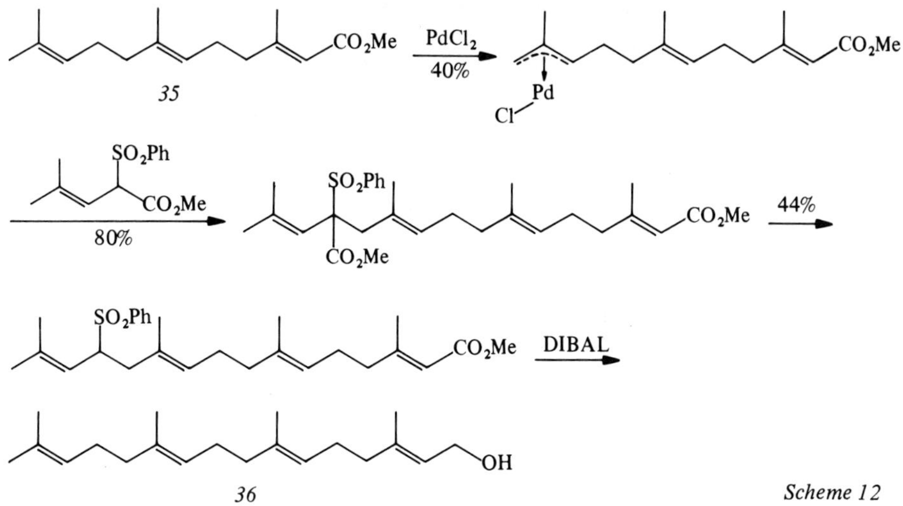

Scheme 12

tion of sulfones *38* with the π-allyl complex *37* derived from 3-methyl-2-butenyl acetate. Reaction of the complex *37* with 3-methyl-1-phenylsulfonyl-5-(2,6,6-trimethyl-1-cyclohexen-1-yl) penta-2,4-diene (*38*) in DMF in the presence of triphenylphosphine gives 1-acetoxy-3,7-dimethyl-5-(phenylsufonyl)-9-(2,6,6-trimethyl-1-cyclohexen-1-yl)nona-2,6,8-triene (*39*) (52% yield), which is converted to vitamin A (*40*)[17] (Scheme 13). This method permits stereocontrolled introduction of C-17 alkyl side

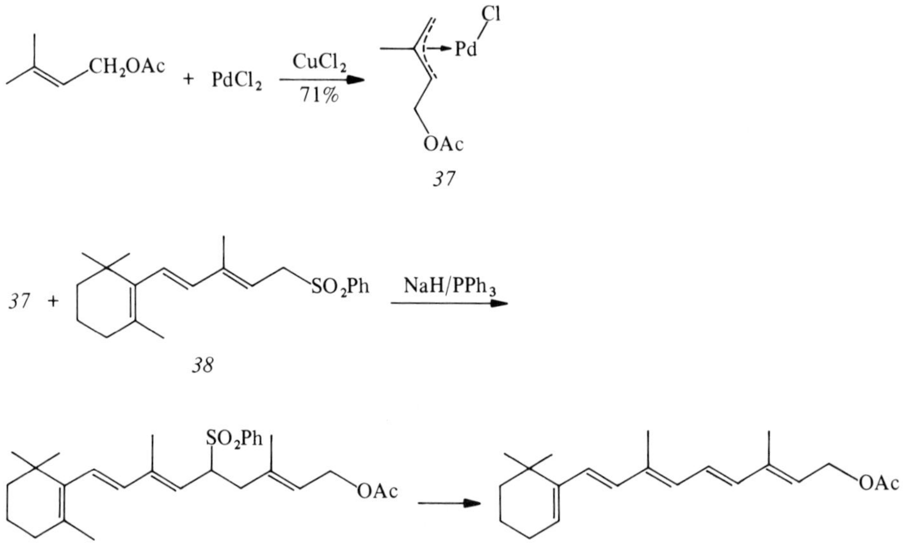

37 *38* *39* *40* *Scheme 13*

chains into steroids. Thus, estrone methyl ether (*41*) is converted to 3-methoxy-19, 24-bis-*nor*-20-isocholane-1,3,5(10),16-tetraenoate (*43*) via the palladium complex *42* and subsequent alkylation of *42* with dimethyl malonate[18] (Scheme 14).

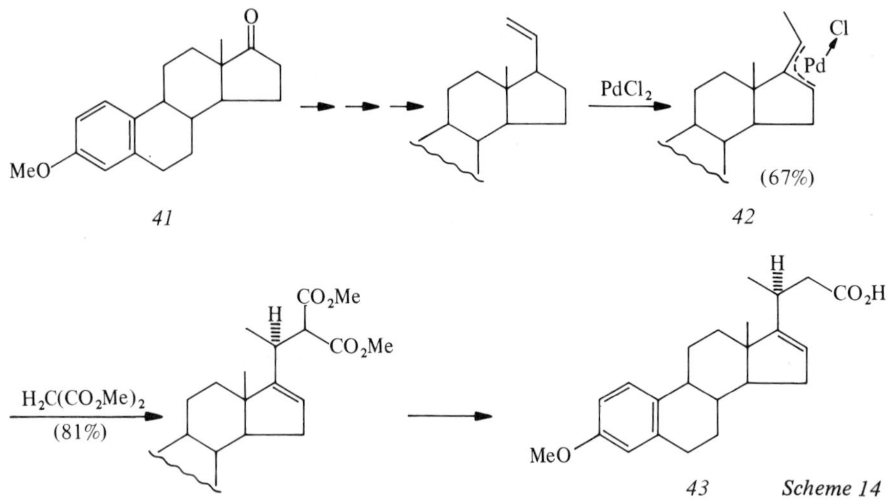

41 *42* *43* *Scheme 14*

2.3 Other Reactions

Transmetallation of organomercury compounds with palladium salts proceeds smoothly to give highly reactive organopalladium compounds which readily react with olefins[19].

$$R{-}Hg{-}X + PdY_2 \longrightarrow [R{-}Pd{-}Y] \xrightarrow{R'{-}CH{=}CH_2} R{-}CH{=}CH{-}R'$$

This method permits vinylation of aromatic compounds. In the isoflavanone synthesis, 4-chromanone (*44*) is converted to the enol ester *45*, which is reacted with phenylpalladium acetate, formed in situ from phenylmercury chloride and $Pd(OAc)_2$, to give – after hydrolysis – isoflavanone (46) (Schema 15). A simple synthesis of

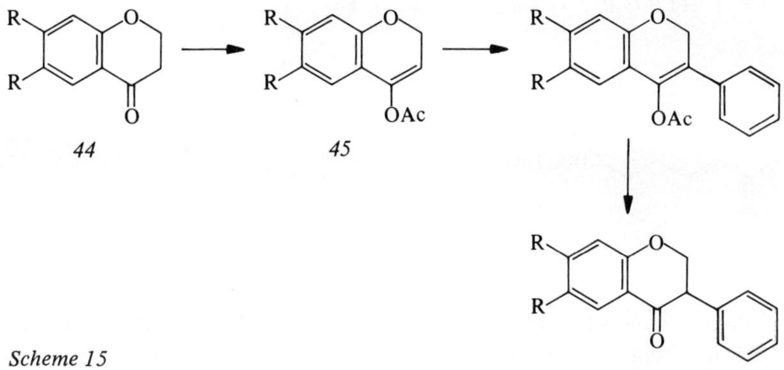

Scheme 15

pterocarpine (*47*) has been achieved by the intramolecular oxypalladation of benzpyrane[21] (Scheme 16). Chloromercuration of propargyl alcohol, followed by car-

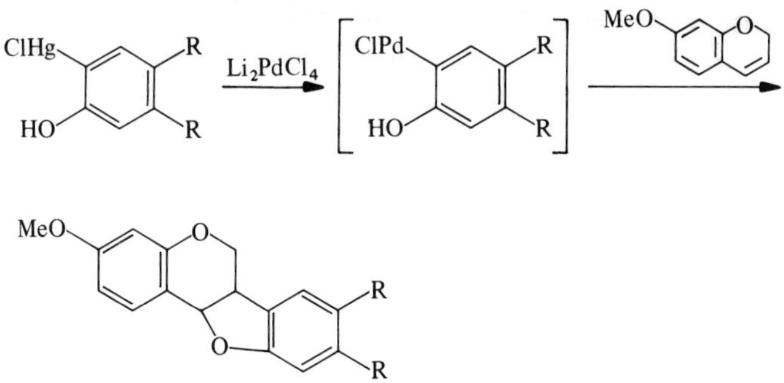

(85%)
47

Scheme 16

bonylation in the presence of $PdCl_2$ affords β-chloro-α-butenolide (*48*) which is converted to β-alkyl-α-butenolides (*49*) by treatment with dialkyllithium cuprate (I)[22]:

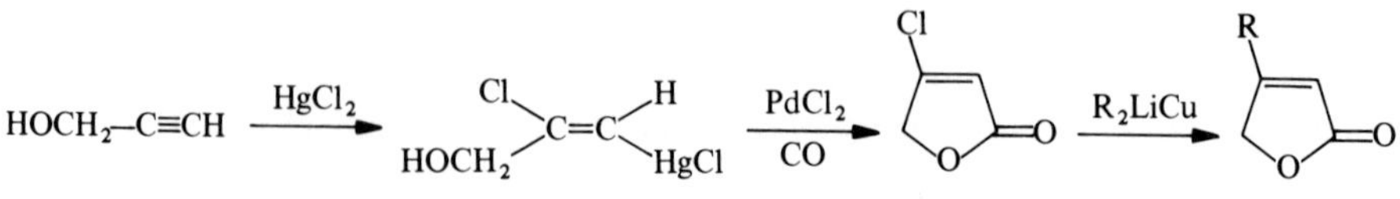

48 49

Due to the chelating effect of the amino group, allylic amines readily undergo nucleophilic addition reaction on their double bond. For example, carbopalladation of allyl-dimethylamine with malonates readily yields a chelating complex. Subsequent olefin insertion of methyl vinyl ketone into this complex gives ω-amino enones[23] (Scheme 17). An interesting application of the facile carbopalladation de-

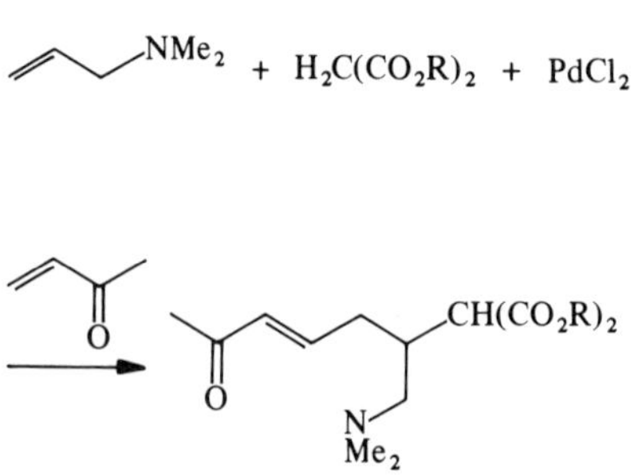

Scheme 17

scribed in Scheme 17 and subsequent olefin insertion is the synthesis of prostaglandin derivatives starting from 2-cyclopentenylamine (*50*)[24]. The key step in this synthesis is the facile introduction of a carbanion and an oxy anion into the cyclopentene ring by virtue of the stabilizing chelating effect of the amino group, followed by olefin insertion into the palladium-carbon σ-bond. Treatment of 2-cyclopentenylamine (*50*) with Li_2PdCl_4 and sodium diethyl malonate in THF at 0 °C for 2 h, followed by the addition of diisopropyl-ethylamine with warming affords diethyl 5-amino-cyclopent-2-en-1-yl malonate (*52*) in 92% yield. This reaction can be explained by the formation of the intermediate carbopalladation product *51*, followed by the elimination of PdH to give *52*. Subsequent treatment of *52* with Li_2PdCl_4, 2-chloroethanol and diisopropyl-ethylamine in DMSO gives rise to the 2-chlroethyloxy substituted aminocyclopentyl-palladium complex *53*, which is immediately treated with *n*-pentyl vinyl ketone. Insertion of this olefin into the Pd-carbon bond leads to the desired enone *54* in 50% yield. This enone has been converted into *55* which is an important intermediate in the synthesis of prostaglandin E and F series (Scheme 18).

3 Catalytic Reactions

Since the reactions described in this chapter proceed with a catalytic amount of palladium compounds they are more useful than the stoichiometric reactions.

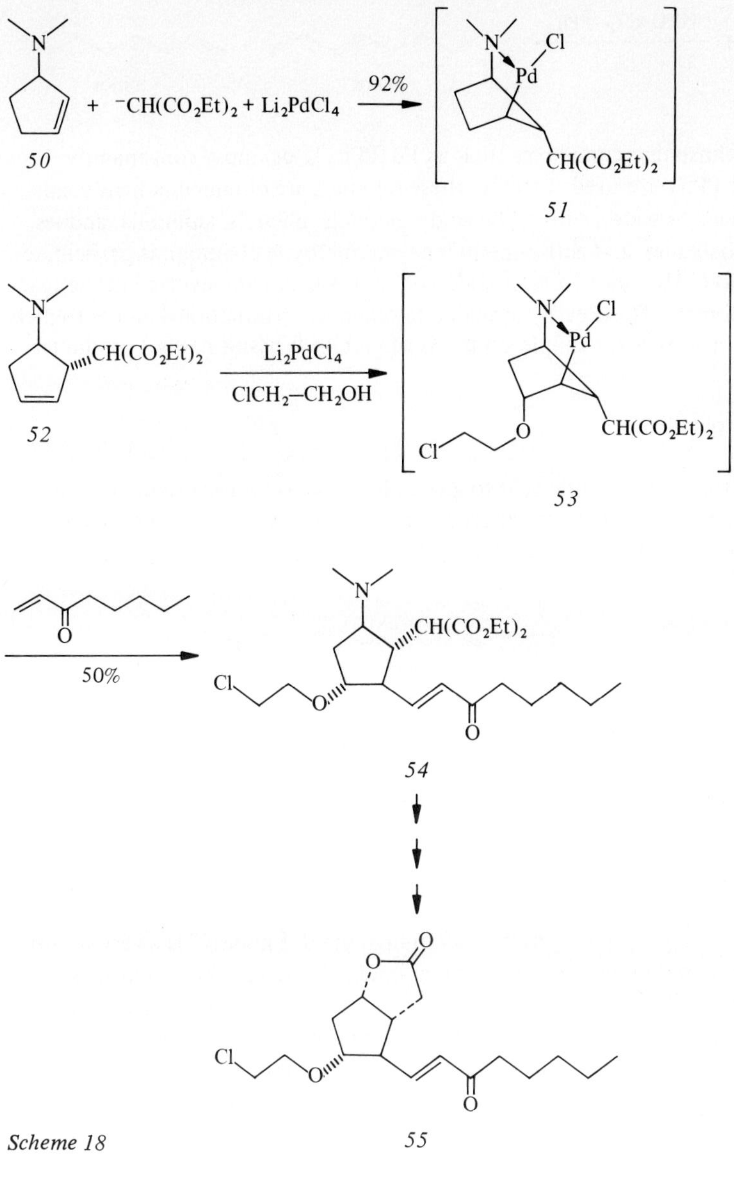

Scheme 18

3.1 Telomerization Reactions of 1,3-Butadiene and Isoprene[2g, h]

Unlike nickel catalysts which form cyclic dimers and trimers (1,5-cyclooctadiene and 1,5,9-cyclododecatriene), palladium compounds catalyze linear dimerization of conjugated dienes. 1,3-Butadiene itself is converted to 1,3,7-octatriene. The reaction most characteristic of palladium is the formation of various telomers. 1,3-Butadiene dimerizes with incorporation of various nucleophiles to form telomers of the following type:

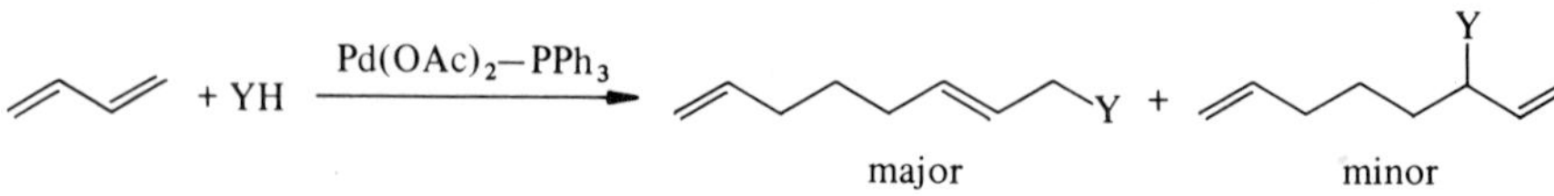

Palladium-phosphine complexes such as Pd $[PPh_3]_4$ or, most conveniently, $Pd(OAc)_2$ and PPh_3 are used. Usually, these telomers are obtained in high yields. Nucleophiles such as water, carboxylic acids, alcohols, phenols, ammonia, amines, enamines, nitroalkanes, and active methylene and methyne compounds participate in telomerization. Also, carbon monoxide and hydrosilanes are involved in the reaction to give telomers. These easily available telomers are trifunctional and extremely useful starting materials for simple synthesis of certain types of natural products.

3.1.1 Acetoxyoctadienes

1,3-Butadiene reacts with acetic acid to give in high yields two acetoxyoctadiene isomers *56* and *57*[25, 26] which are interconvertible in the presence of the palladium catalyst (Scheme 19). The telomer *57* is hydrolyzed to 1,7-octadien-3-ol (*58*) which

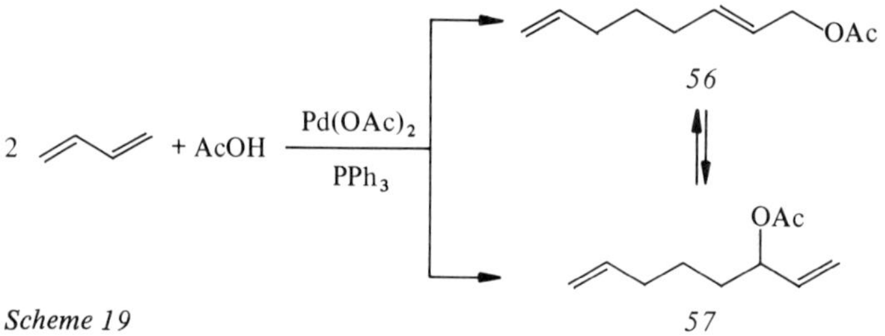

Scheme 19

is oxidized to the corresponding ketone *59* in high yield. Enone *59* is a very useful reagent for bisannulation because its terminal double bond may be regarded as a masked ketone which can be readily unmasked on treatment with oxygen in the presence of $PdCl_2/CuCl$ to form – after *Michael* addition at the enone moiety of *59* – the 1,5-diketone *60* (Scheme 20). The enone *59* is the cheapest and most readily available bisannulation reagent, permitting a simple total synthesis of steroids[27]. Thus, in the simplest example, Michael addition of the enone *59* to cyclohexanone

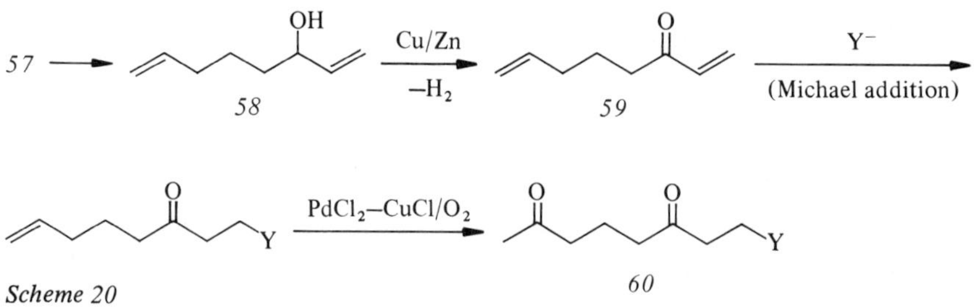

Scheme 20

enamine (*61*) in boiling dioxane, followed by aldol condensation, yields 4-(3-butenyl)-3-oxo-Δ^4-octalin (*62*). The terminal double bond is then oxidized to the ketone *63* by $PdCl_2/CuCl/O_2$; subsequent aldol condensation using t-$C_5H_{11}OK$ leads to the tricyclic ketone *64* (Scheme 21).

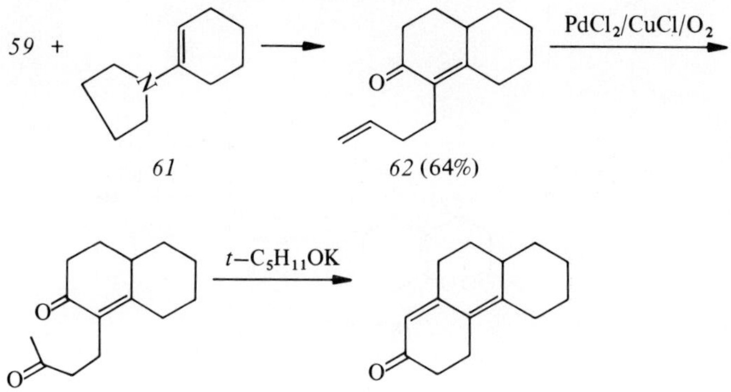

61 *62* (64%) *63* (83%) *64* (73%)

Scheme 21

The important intermediate *66* of the steroid synthesis has been prepared by the application of the same reaction sequence to 2-methyl-1,3-cyclohexanedione (*65*) (Scheme 22). A synthesis of (+)-19-nortestosterone (*69*) starts with the Michael addition of the optically active oxo ester *67* to 1,7-octadien-3-one (*59*) catalyzed by sodium hydride, the ester group being removed by heating in aqueous HMPA with sodium iodide to give the dione *68*. The aldol condensation catalyzed by sodium hydroxide proceeds in 90% yield. The terminal double bond is oxidized with $PdCl_2/CuCl$ to the methyl ketone and the internal olefinic double bond subsequently hydrogenated. The final reaction step involves aldol condensation in refluxing

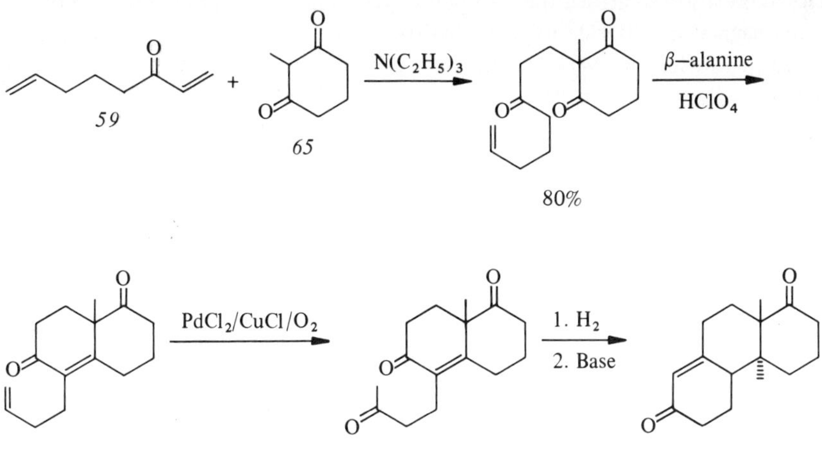

59 *65* 80% 73% 75% *66* (92%)

Scheme 22

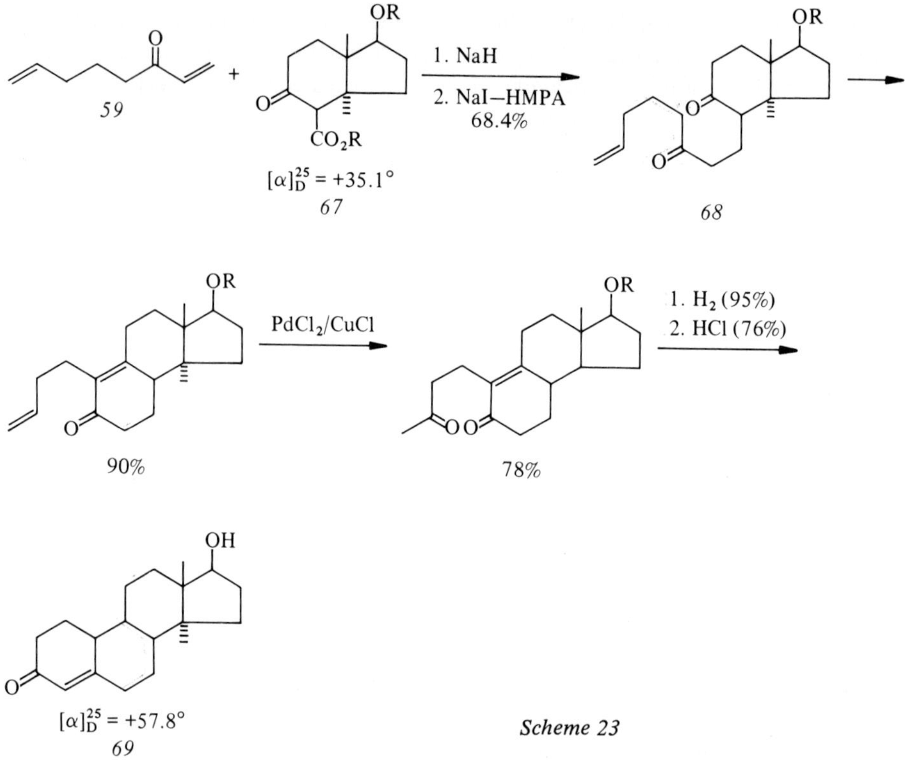

Scheme 23

methanol containing hydrogen chloride to afford (+)-19-nortestosterone (*69*) (Scheme 23). The enone *59* is converted to the trisannulation reagent[28] *72* as follows: Addition of malonate diester to *59* to yield the 3-oxo-7-octenylmalonate diester *70* which is reduced to the corresponding 3-hydroxy derivative; hydrolysis and decarboxylation to afford the δ-(3-pentenyl)-δ-valerolactone *71*; reaction of *71* with vinylmagnesium chloride to give 7-hydroxy-1,11-dodecadien-3-one which is acetylated to *72* (Scheme 24) . The *Michael* addition of the reagent *72* to 2-methylcyclohexa-1,3-dione (*65*) is followed by aldol condensation and reduction of one of

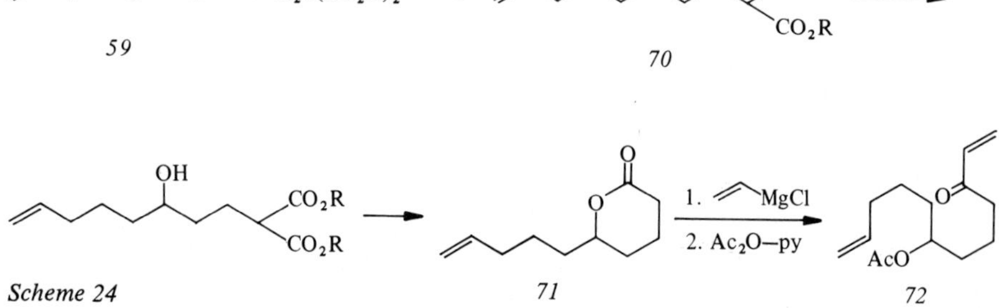

Scheme 24

the carbonyl groups with $NaBH_4$ to give *73*. The olefinic double bond of the enone system of *73* is reduced with lithium in ammonia to give the *trans*-fused CD ring system. Hydrolysis of the ester and oxidation with chromic oxide afford the trione *74* which is cyclized to the tricyclic ketone *75*. The unmasking of the terminal olefinic double bond and hydrogenation yield the 1,5-diketone *76* which is subjected to intramolecular aldol condensation to give *D*-homo-4-androstene-3,17a-dione (*77*) (Scheme 25).

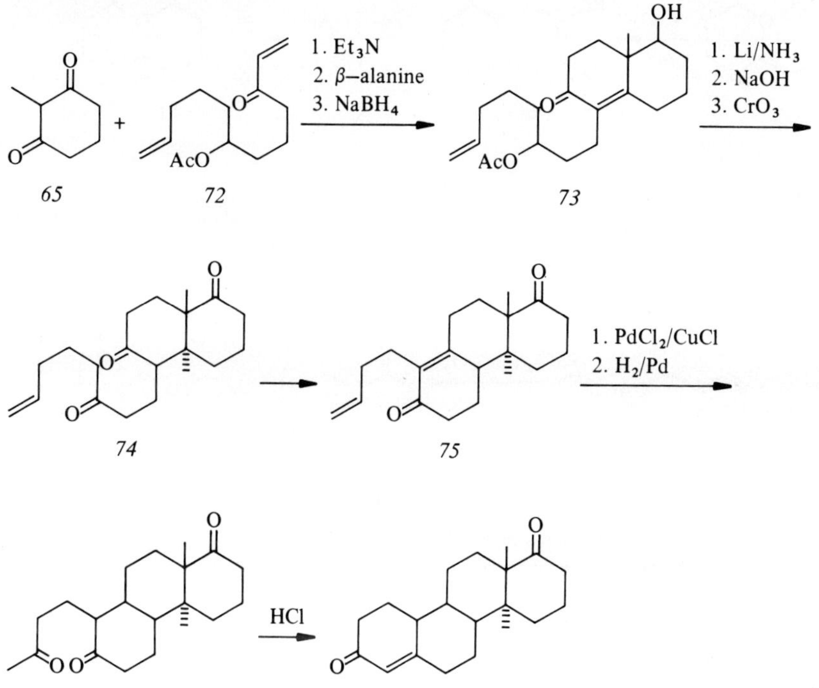

Scheme 25

1,3-Butadiene telomers are good starting materials for macrolide synthesis. Macrolide syntheses based on intramolecular carbon-carbon bond formation using 1,3-butadiene telomers as starting materials have been carried out. Comparison of *56* with *81* suggests that the acetate *56* has suitable functionality for the synthesis of diplodialide B (*81*). The steps involve conversion of the acetate *56* to the tetrahydropyranyl ether, oxidation of the terminal olefin to the methyl ketone *78* with $PdCl_2/CuCl/O_2$ and subsequent reduction to the corresponding alcohol with $NaBH_4$. Reaction of this alcohol with bromoacetyl bromide yields the bromoacetate *79*. The tetrahydropyranyl moiety is then removed and the liberated alcohol oxidized with CrO_3 to the aldehyde *80*. The final step is the intramolecular Reformatsky reaction to form diplodialide B (*81*)[29]. This cyclization is promoted by zinc in the presence of $AlEt_2Cl$ as an activator[30]. In this reaction, aluminum protects the hydroxy group from dehydration (Scheme 26). Other ten-membered lactones can also be conveniently prepared from 2,7-octadienyl acetate (*56*)[31]. Again the terminal

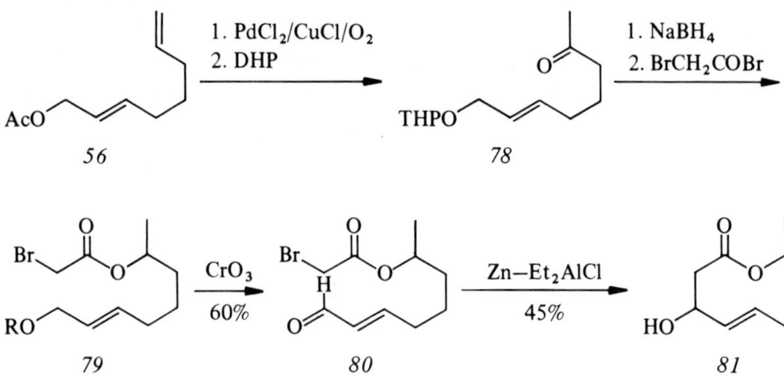

Scheme 26

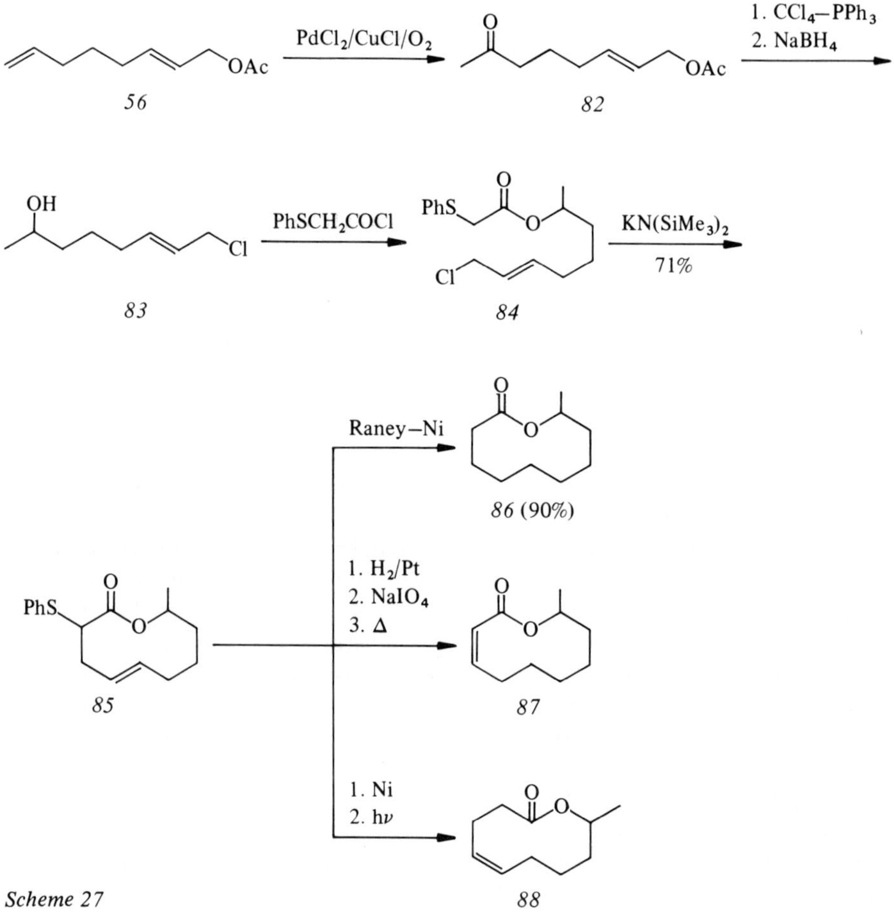

Scheme 27

double bond in *56* is oxidized to the methyl ketone *82* which is reduced to the corresponding alcohol. Subsequent hydrolysis of the acetoxy group and treatment with CCl_4/PPh_3 yield the allylic chloride *83*. Treatment of *83* with phenylthioacetyl chloride leads to the ester *84*. Intramolecular alkylation to the unsaturated lactone *85* is carried out by adding a solution of *84* to a THF solution of two equivalents of $KN[Si(Me_3)]_2$ as a base at reflux over a period of 2 h. Since this base is thermally stable and displays low nucleophilicity, it is most suitable for this cyclization, the ester group remaining unaffected. The reaction proceeds rapidly and hence high dilution is unnecessary. Treatment of *85* with Raney-nickel affords 9-decanolide (*86*) which is a natural product isolated from *phoracantha synonyma*[32]. Hydrogenation and subsequent oxidative removal of phenylthio group provides the unsaturated lactone *87* which has been prepared as a precursor of diplodialide C[33]. The phenylthio group of *85* can also be removed by treatment with deactivated Raney nickel without attacking the double bond. Photochemical *trans-cis* isomerization to form phoracatholide J (*88*) is a well-known reaction[34] (Scheme 27).

1-Chloro-7-hydroxy-2-octene (*83*) is a convenient starting material for the ring forming reaction of lasiodiplodin *92* (Scheme 28). In the first step, the ester *90* is

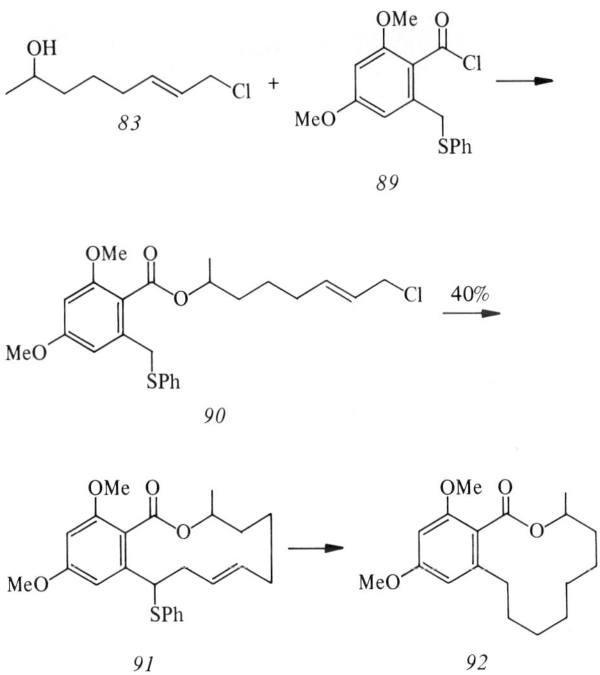

Scheme 28

prepared from *83* and *89*. Intramolecular alkylation of the anion generated upon treatment of *90* with $KN[Si(CH_3)_3]_2$ in refluxing THF affords the unsaturated lactone *91*. Reduction of the double bond and removal of the phenylthio group are achieved by treatment of *91* with Raney-nickel to give the dimethyl ether of lasiodiplodin (*92*)[35]. Zearalenone is another type of orsellinic acid type macrolide and

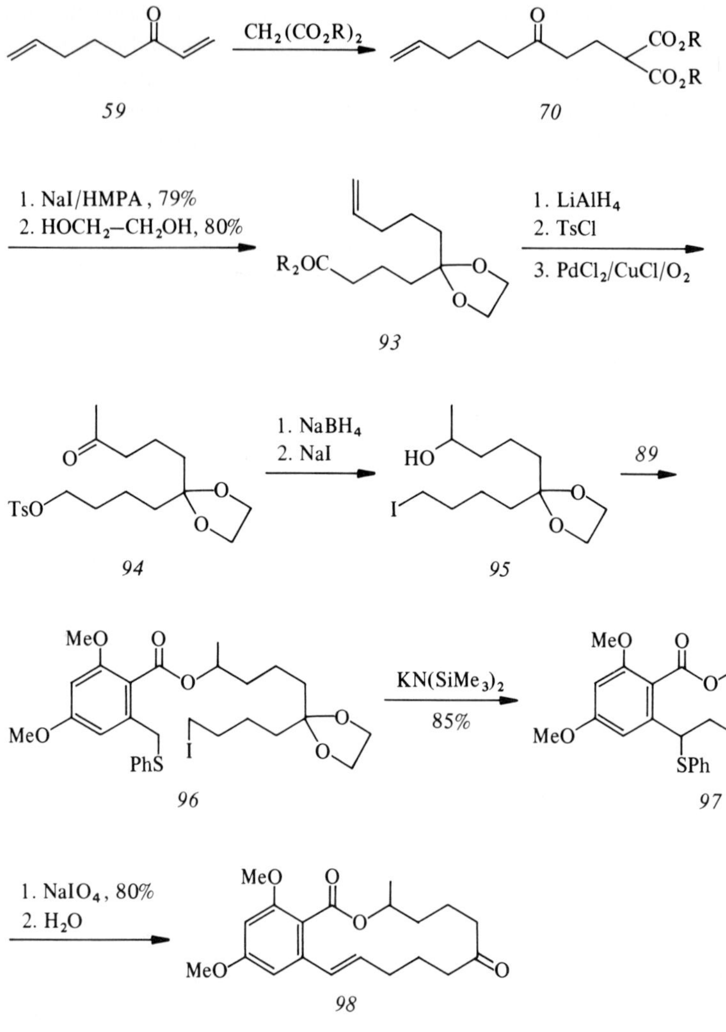

Scheme 29

its ring is easily prepared from the enone *59* by the following reaction sequence[36] (Scheme 29). The enone *59* is subjected to the Michael reaction with dialkyl malonate to give the oxodiester *70*. One of the ester groups is cleaved by heating in HMPA and the carbonyl group protected as the ethylenedioxy derivative *93*. The ester is reduced to the alcohol which is converted to the tosylate. The terminal olefinic double bond is then oxidized with $PdCl_2/CuCl/O_2$ to give the methyl ketone *94*. Reduction of this ketone and displacement of the tosyloxy group with sodium iodide yield *95* which is esterified with *89* to give the corresponding ester *96*. Subsequent cyclization is carried out with three equivalents of the base in refluxing THF for 2 h to give the lactone *97*. Oxidative removal of the phenylthio group and deacetalization affords zearalenone dimethyl ether (*98*). The simplest natural product prepared from the telomer *56* or *57* is 1-octene-3-ol (*100*) (Scheme 30) (*Matsutake* alcohol), a fragrant compound contained in a Japanese mushroom. The synthesis has been accomplished by two methods. In one method, the terminal double bond of *56* is hydrogenated selectively using, dichloro-tris-(triphenylphosphine)-ruthenium(II) $RuCl_2[PPh_3]_3$, as a catalyst. Subsequent allylic rearrangement catalyzed by $Pd(OAc)_2/PPh_3$ produces the acetate of 1-octen-3-ol (*99*)[37]. In the second route[38], highly selective reduction of the terminal double bond at C_7 in *57* without attacking the terminal double bond at C_1 is carried out by hydroalumination reaction catalyzed by a titanium catalyst. After hydrolysis of the acetoxy group the resultant hydroxy group is protected as diisobutylaluminum alkoxide by treatment with diisobutylaluminum hydride. This protecting group also blocks the neighboring terminal double bond. Then $LiAlH_4$ is added in the presence of bis-cyclopentadienyl-dichloro-titanium(IV) as a catalyst. Hydrolysis of the resultant product gives 1-octen-3-ol (*100*) with high selectivity.

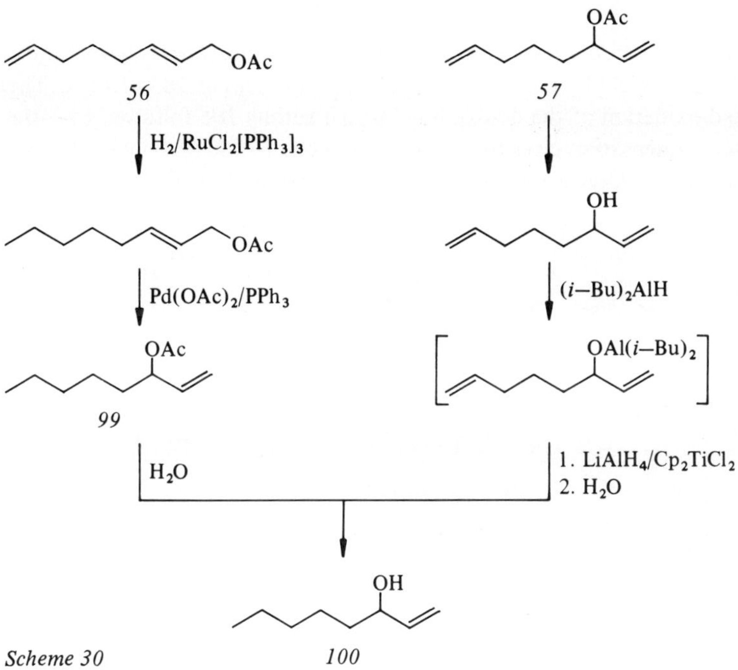

Scheme 30

1,7-Octadien-3-ol (*58*) is a suitable starting material for the synthesis of lipoic acid (*105*) (Scheme 31) without changing the C-atom number[39]. Hydroboration of the two double bonds in *58* produces the triol *101*. The 1,3-diol moiety is protected by acetal formation *102* and the terminal free hydroxy group oxidized with CrO_3 to yield the carboxylic acid *103*. Treatment of *103* with thiourea/HI leads to the 1,3-dithiol *104*. $FeCl_3$ catalyzed oxidation of this dithiol affords lipoic acid (*105*).

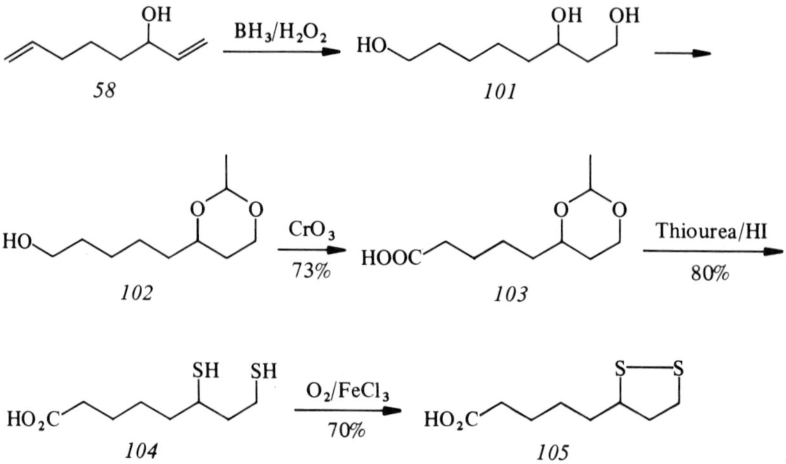

Scheme 31

Vinyl 2-octenyl ether *106*, obtained from 2,7-octadienyl acetate (*56*) is converted to the aldehyde *107* in 79% yield by [3,3] sigmatropic rearrangement at 183–190 °C. Then $PdCl_2$-catalyzed oxidation of the double bond to the ketone *108* followed by intramolecular aldol condensation gives the 2-cyclopentenone *109* in 68% yield, which is converted to methyl dihydrojasmonate (*110*)[40]. Treatment of allyl 2-octenyl

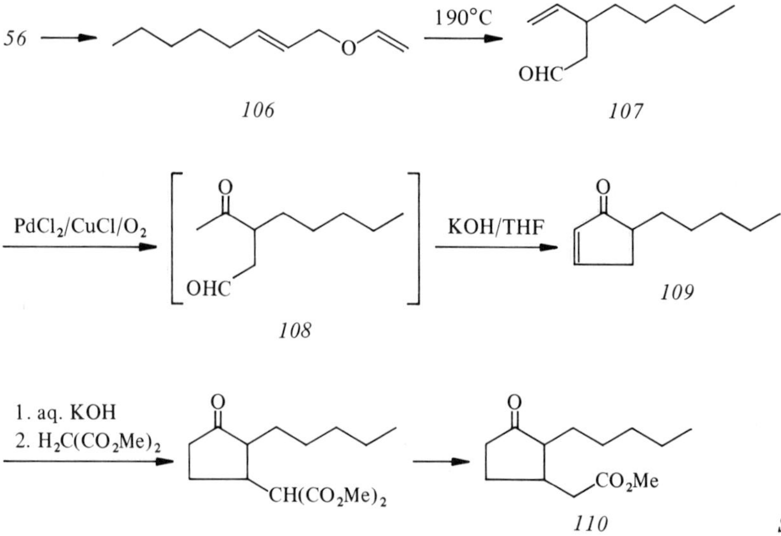

Scheme 32

ether *111* with $RuCl_2[PPh_3]_3$ as a catalyst at 200 °C causes migration of the terminal double bond and subsequent [3,3] sigmatropic rearrangements to give the aldehyde *112* in 61% yield[41]. Then $PdCl_2$ catalyzed oxidation to the oxo aldehyde *113* and intramolecular aldol condensation affords dihydrojasmone *10*[40] (Scheme 33). 2,15-

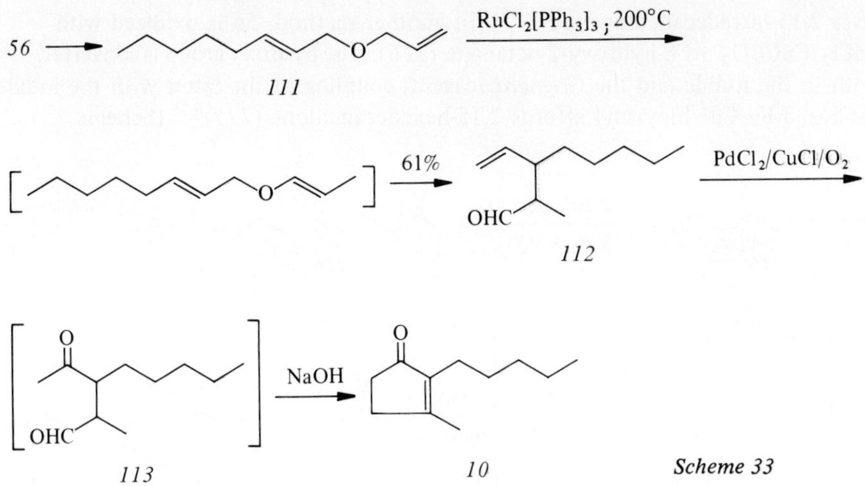

Scheme 33

Hexadecanedione (*115*), from which muscone has been prepared by intramolecular aldol condensation[42], is synthesized by C–C coupling of a mixture of 1-chloro-2,7-octadiene and 3-chloro-1,7-octadiene obtained from the corresponding acetoxyoctadienes *56* and *57*, respectively (Scheme 34). This allylic coupling promoted by iron powder yields

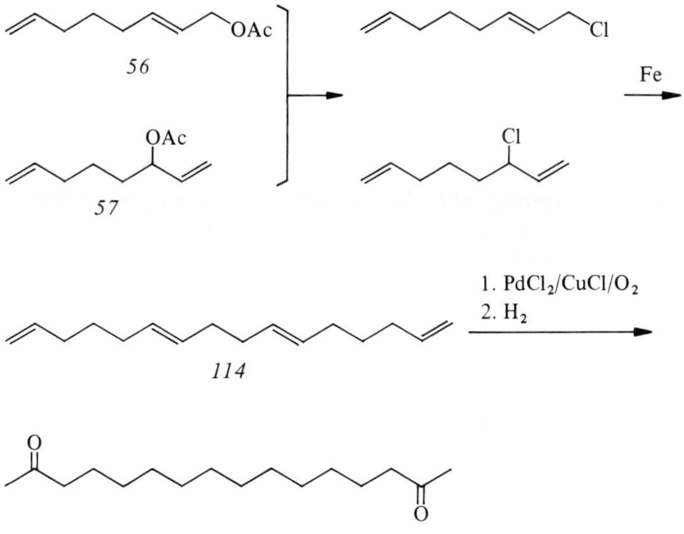

Scheme 34

a mixture of linear and branched 16-carbon tetraenes. The linear hexadecatetraene *114* as a main product is separated by distillation from the isomers. The oxidation of the terminal double bonds of this tetraene with the catalyst system $PdCl_2/CuCl$ followed by hydrogenation of the remaining internal double bonds gives 2,15-hexadecanedione (*115*)[43]. In another method, *56* is oxidized with $PdCl_2/CuCl/O_2$ to 8-hydroxy-2-octanone (*116*). The hydroxy group is converted both to the iodide and the *Grignard* reagent; coupling of the latter with the iodide catalyzed by CuI/bipyridyl affords 2,15-hexadecanedione (*115*)[44] (Scheme 35).

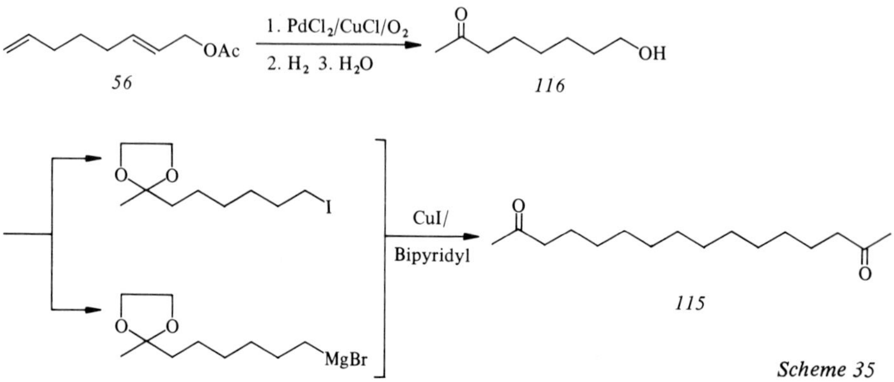

Scheme 35

3.1.2 Octadienyl Phenyl Ethers

Phenol reacts with butadiene to give the following telomers[45]:

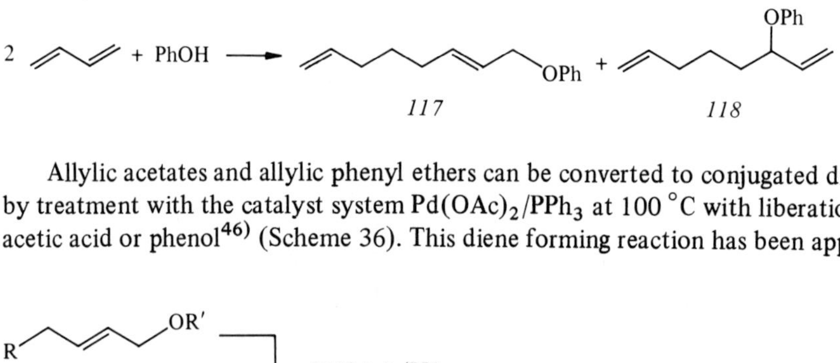

Allylic acetates and allylic phenyl ethers can be converted to conjugated dienes by treatment with the catalyst system $Pd(OAc)_2/PPh_3$ at 100 °C with liberation of acetic acid or phenol[46] (Scheme 36). This diene forming reaction has been applied

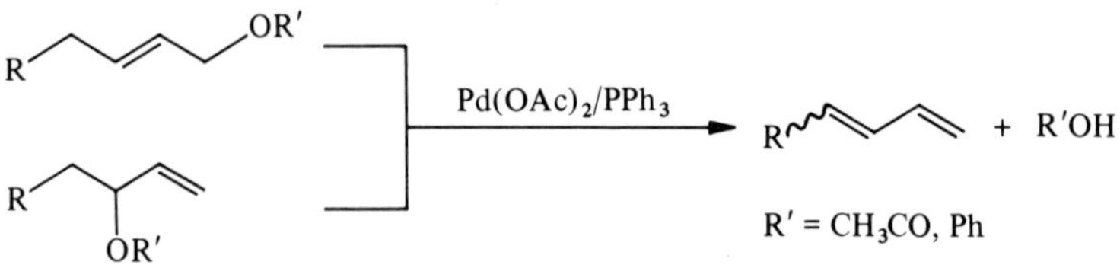

Scheme 36

to the synthesis of 12-acetoxy-1,3-dodecadiene (*123*), a pheromone of *Diparopsis castanea*[47] (Scheme 37): 8-Phenoxy-1,6-octadiene (*117*) is converted to the alcohol *119a* by selective hydroboration of the terminal double bond and then tosylated. Treatment of the tosylate with sodium iodide yields the iodide *119b*. This iodide is coupled with the *Grignard* reagent *120* of 4-chlorobutanol (prepared from THF) using CuI/2,2′-bipyridyl as a catalyst. Then the phenyl ether *121* is converted to the

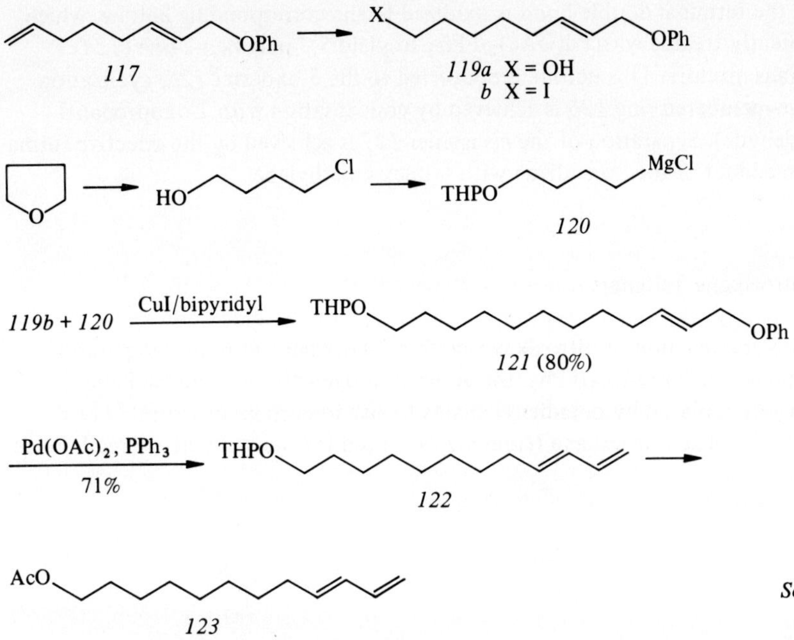

Scheme 37

conjugated diene *122* with $Pd(OAc)_2/PPh_3$ in 71% yield. Acetylation of *122* completes the synthesis of the pheromone *123*. The diene is a mixture of cis/trans isomers in a ratio of 36:64. The method of diene synthesis has been applied to the synthesis of pyrethrolone *127*[48] (Scheme 38). Starting from the phenyl ether telomers *117*

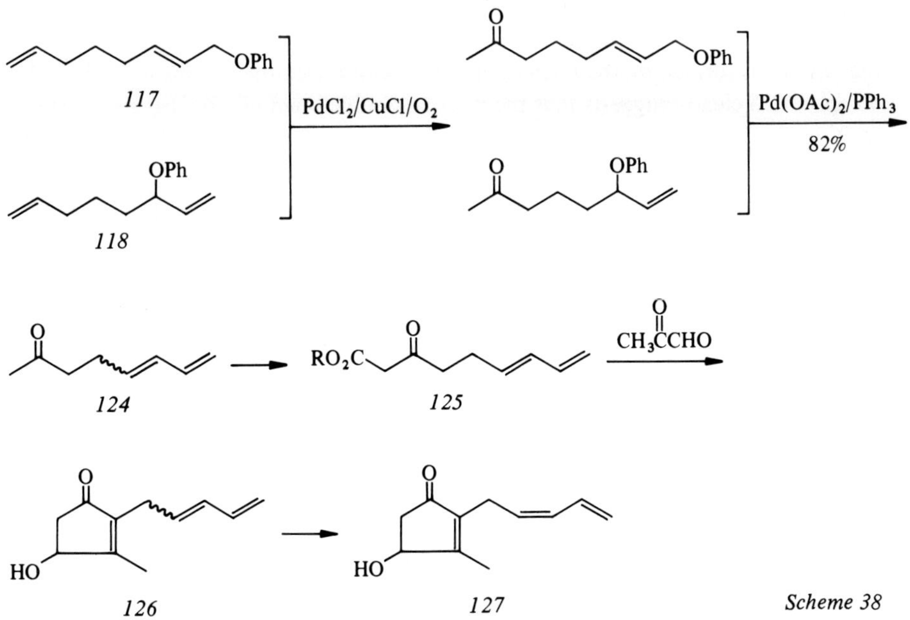

Scheme 38

and *118*, the terminal double bond is oxidized to the corresponding ketone, which is subsequently treated with $Pd(OAc)_2/PPh_3$ to yield 5,7-octadien-2-one (*124*) as a cis-trans mixture. This ketone is converted to the 3-oxoester *125;* cyclization to the five-membered ring *126* is achieved by condensation with 2-oxopropanal (pyruvaldehyde). Separation of the *cis* isomer *127* is achieved by the selective formation of an adduct of the *trans* diene with tetracyanoethylene.

3.1.3 Nitroalkane Telomers

In the catalytic reaction of nitroalkanes with 1,3-butadiene at room temperature in the presence of $Pd(OAc)_2/PPh_3$, the active hydrogen atoms of nitroalkanes are stepwisely replaced by octadienyl chains to give long-chain unsaturated nitro compounds[49]. The nitroethane telomer *129* is used for the synthesis of recifeiolide

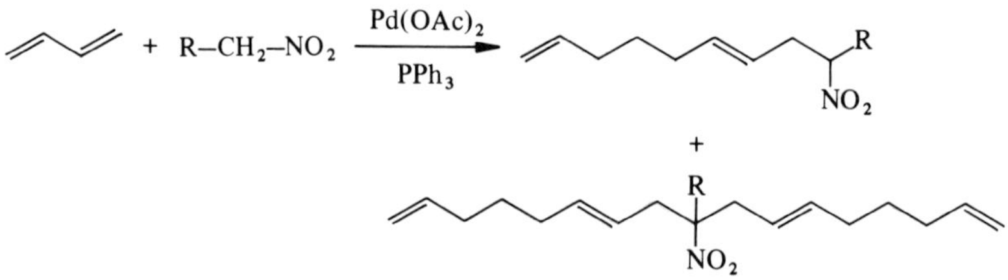

134, a naturally occurring 12-membered lactone[31, 50]. The usefulness of the compound *129* as a starting material for simple synthesis of recifeiolide is apparent by the comparison of *129* with *134* (Scheme 39). The mechanistic consideration that the telomer is formed by the nucleophilic attack of nitroethane at bis-π-allylpalladium complex *128* clearly suggests that the internal double bond of *130* is at the right posi-

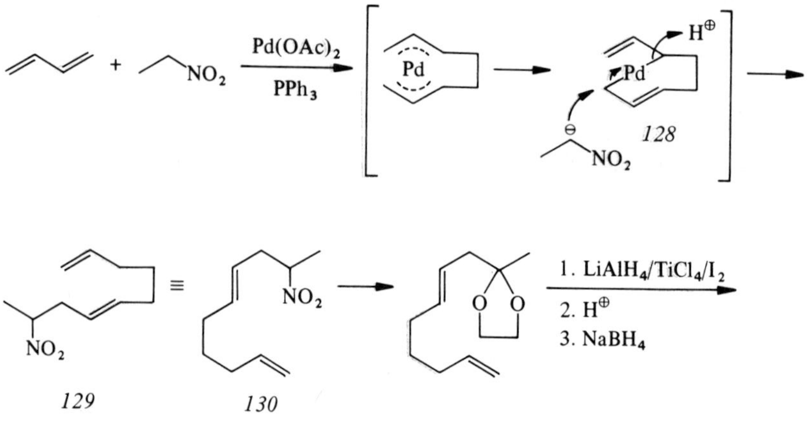

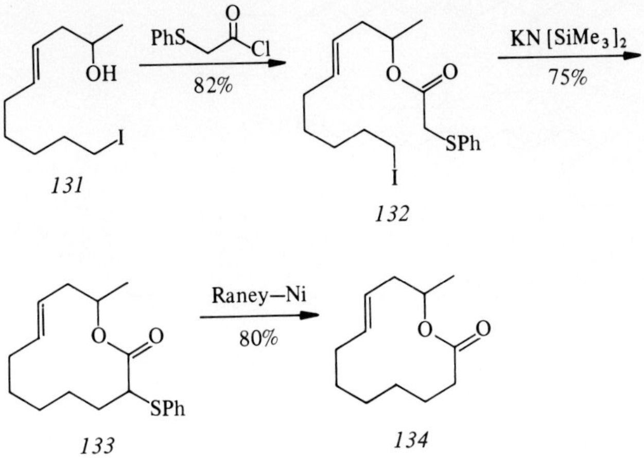

Scheme 39

tion possessing the required *trans* configuration as in recifeiolide. The nitro group is converted to the ketone by the *Nef* reaction (MeONa/$TiCl_4$), and the carbonyl group protected by acetalization. The terminal double bond is selectively hydroaluminated with $LiAlH_4$ using $TiCl_4$ and subsequently quenched with iodine to give the iodide. The protected carbonyl group is then liberated and reduced to the alcohol *131*. Treatment of *131* with phenylthioacetyl chloride affords the corresponding phenylthioacetate *132*. The intramolecular alkylation of *132* using $KN[Si(CH_3)_3]_2$ gives the lactone *133* in 75% yield. The phenylthio group is removed upon treatment with deactivated Raney nickel to give recifeiolide *134* in 80% yield. 9-Nitro-1,6,11,16-heptadecatetraene (*135*) is one of the telomerization products of 1,3-butadiene and nitromethane. The telomer *135* has a linear 17-carbon atom chain with the nitro group at its center. Civetonedicarboxylic acid (*136*), a precursor of civetone, has

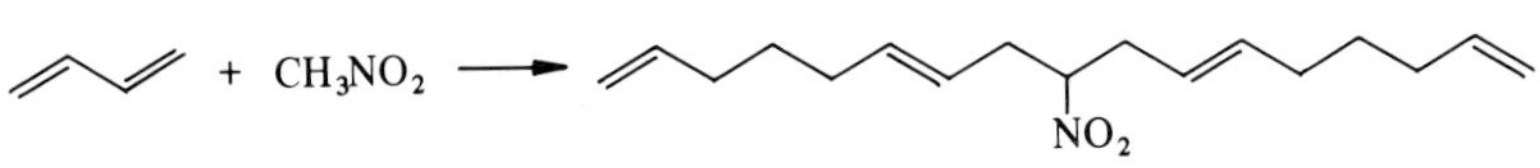

135

been prepared according to Scheme 40 without changing the carbon atom numbers. The nitro group is converted to the ketone by the *Nef* reaction (MeONa/H_2SO_4); subsequently, the carbonyl group is protected as acetal. The terminal double bonds are selectively hydroborated with 9-BBN and the internal double bonds hydrogenated. Then the terminal hydroxy groups are oxidized to the carboxy groups to afford civetonedicarboxylic acid (*136*). Dimethyl civetonedicarboxylate (*137*) is cyclized by acyloin condensation to give the acyloin *138* in 65% yield. The ketol is oxidized to the α-diketone and then converted to dihydrazone *139* which is oxidized to the cyclic alkyne *140* with CuCl in pyridine. Hydrogenation catalyzed by the *Lindlar* catalyst affords cis-civetone (*141*)[51] (Scheme 40).

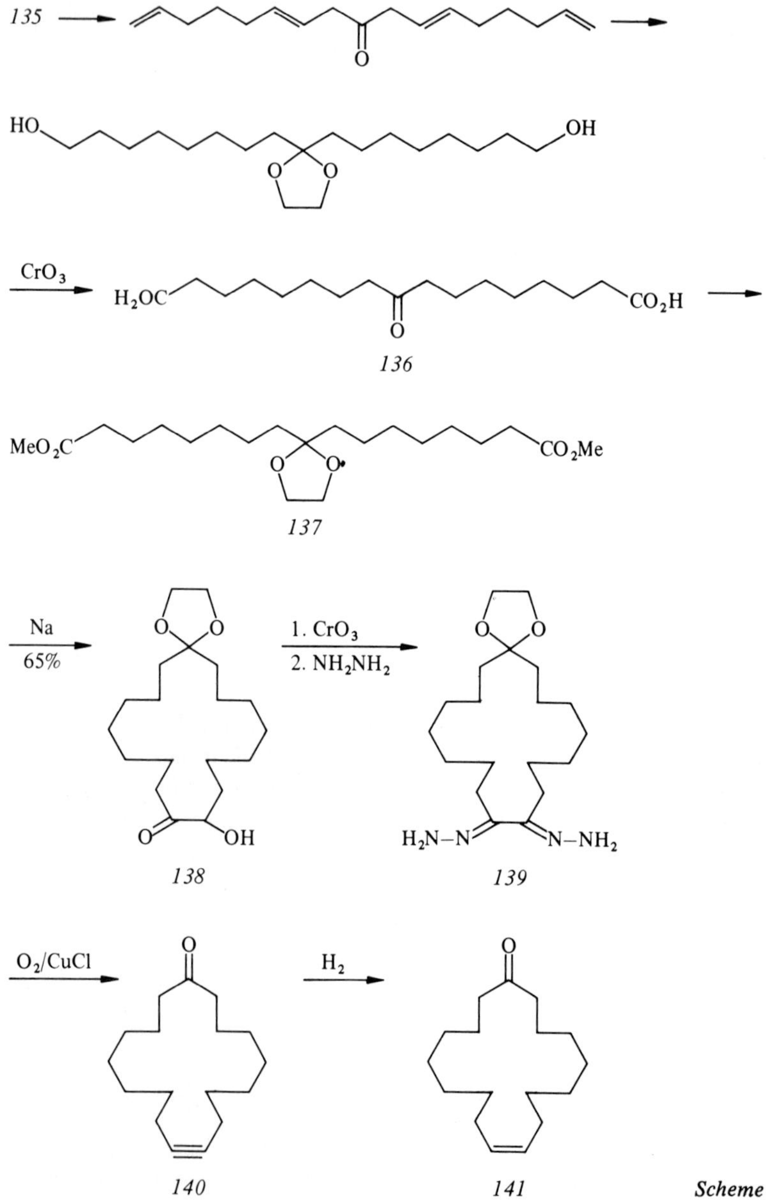

Scheme 40

3.1.4 3,8-Nonadienoate

Carbonylation of butadiene in alcohol catalyzed by $Pd(OAc)_2/PPh_3$ affords 3,8-nonadienoate *142* in high yield[52, 53]. The synthesis of 2-decenedioic acid (*144*) (royal jelly acid) may be carried out as follows[54]: Carbonylation of *142* in alcohol

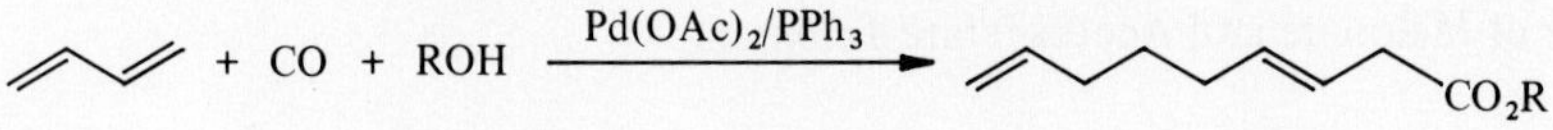

142

using $Co_2(CO)_8$/pyridine as a catalyst yields the linear diester *143* as the main product (62% yield, selectivity 80%). Treatment of *143* with concentrated alcoholic KOH causes hydrolysis and concomitant double bond migration to give royal jelly

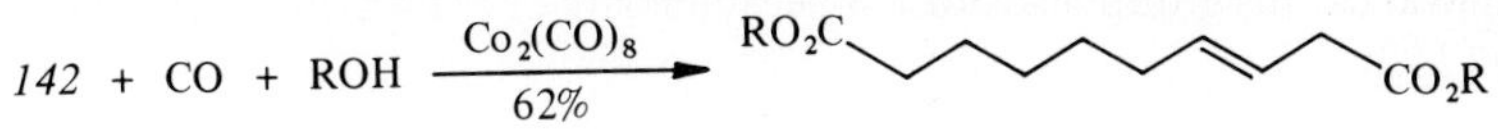

143

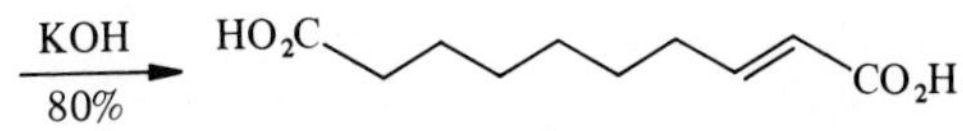

144

acid (*144*) as a crystalline compound. The terminal double bond of *142* is oxidized with $PdCl_2/CuCl/O_2$ to 8-oxo-3-nonenoate *145*. Hydrogenation of the double bond and subsequent hydrolysis of the ester group in *145* afford 8-oxononanoic acid (*146*) as crystals. *Kolbe* electrolysis of *146* yields crystalline 2,15-hexadecanedione (*115*) in 63% yield[55]. Formation of the cyclic α,β-unsaturated ketone *147* from *115* in 17% has been reported[42] (Scheme 41). The yield is greatly improved by using dialkylaluminum phenoxide/tertiary amine as an efficient reagent for the regioselective aldol condensation of 2,15-hexadecanedione at the methyl side. The cyclized ketone *147* is isolated in 65% yield (78% based on consumed *115*) by using di-iso-butylaluminum phenoxide and pyridine as the condensing agent. Hydrogenation of *147* affords muscone (*148*)[56] (Scheme 41).

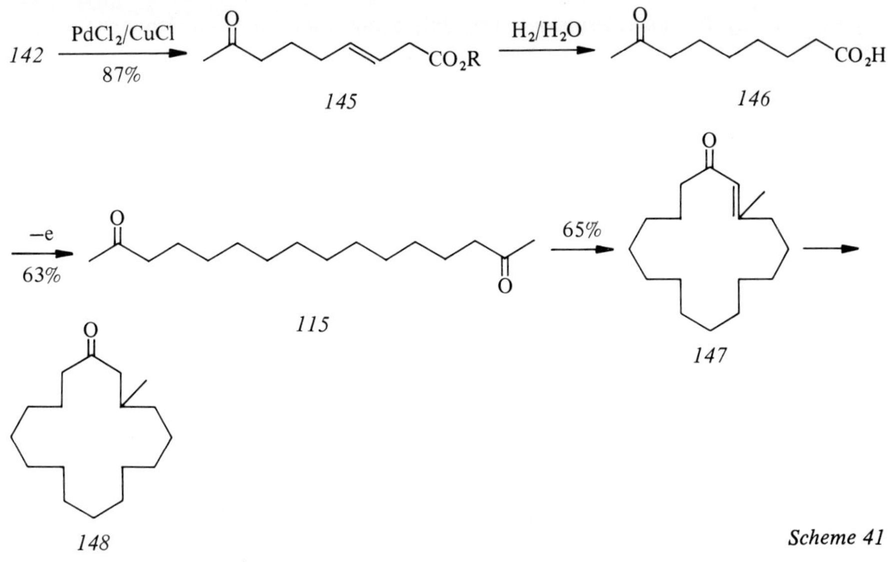

Scheme 41

3.1.5 Use of Malonate and Acetoacetate Telomers

The 1:2 telomer of malonate and butadiene *149* is another useful compound[57]. The first example is the synthesis of pellitorine (*152*) (Scheme 42), a naturally occurring pesticide[58]. The terminal double bond in *149* is hydrogenated selectively using dichloro-tris-(triphenylphosphine)-ruthenium (II) as a catalyst. Partial hydrolysis affords the potassium salt of the monoester *150* which is treated with diphenyl diselenide to displace one of the carboxyl groups by the phenylselenenyl group. Oxidative removal of this group leads to 2,4-decadienoate (*151*) which is converted to pellitorine (*152*).

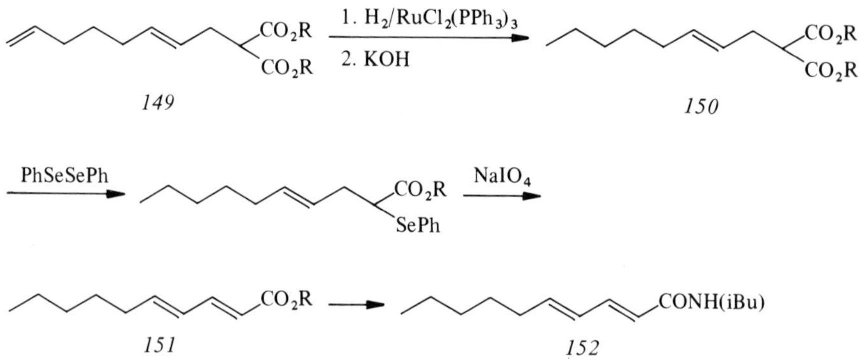

Scheme 42

One of royal jelly acids (10-hydroxy-2-decenoic acid) (*154*) is prepared from the telomer of acetoacetate *153*[59] (Scheme 43). Treatment of *153* with sodium ethoxide and hydroboration lead to 10-hydroxy-4-decenoate. Then the internal double bond is reduced and then reintroduced at the conjugated position by the addition of phenylselenenyl bromide. Subsequent oxidative removal of the phenylthio group yields *154*.

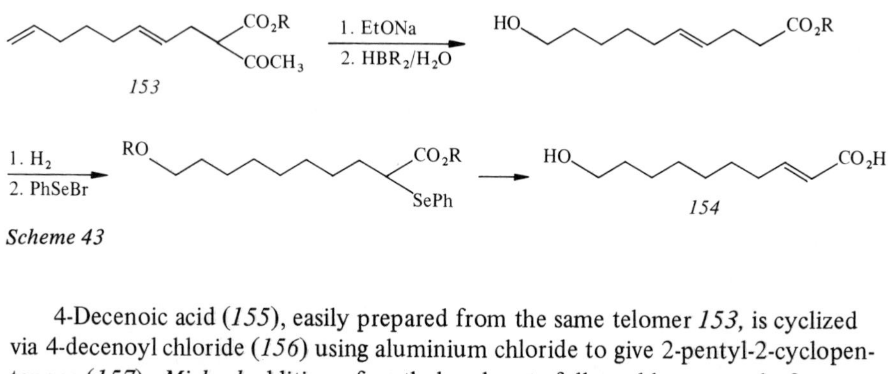

Scheme 43

4-Decenoic acid (*155*), easily prepared from the same telomer *153*, is cyclized via 4-decenoyl chloride (*156*) using aluminium chloride to give 2-pentyl-2-cyclopentenone (*157*). *Michael* addition of methyl malonate followed by removal of one ester group affords methyl dihydrojasmonate (*110*)[60] (Scheme 44).

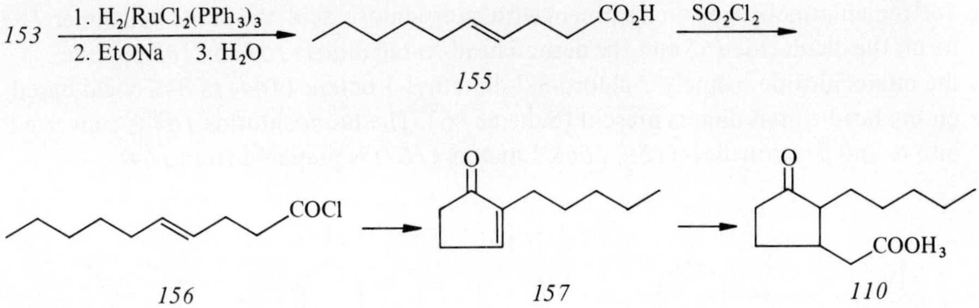

Scheme 44

3.1.6 Dimerization of Isoprene

So far, selective dimerization of isoprene with a palladium catalyst to form natural type terpenoids has not been achieved. The reaction of isoprene with methanol under certain conditions results in a head-to-tail addition product (*158*). However, the methoxy group is introduced at a position differing from that of the oxygen-containing group in natural products[61]. This telomer is converted to citronellol (*159*) by the sequence of reactions described in Scheme 45[62].

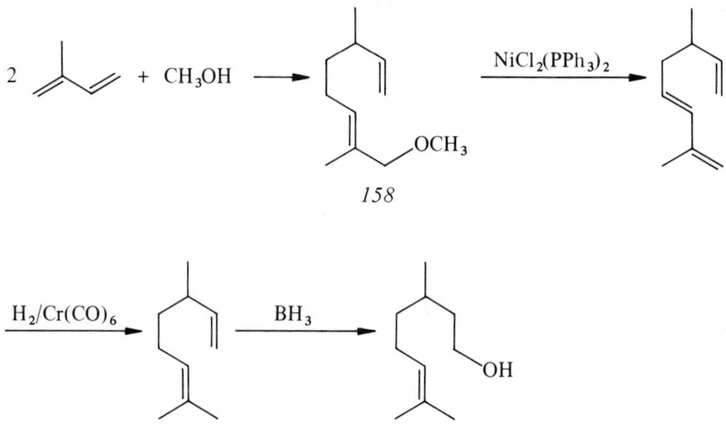

Scheme 45

The reductive dimerization of isoprene in formic acid in the presence of triethylamine at room temperature using a 1% palladium phosphine catalyst gives dimers in up to 79% yield[63]. Higher selectivity with respect to head-to-tail dimer is obtained by using a 1:1 ratio of $Pd(OAc)_2$ to arylphosphines. The use of THF as a solvent causes a favorable effect. By a scaled-up reaction with 0.5 mol of isoprene using π-allylpalladium acetate and *o*-tolyphosphine, the overall yield of the dimers is 87%; these containing 71% of head-to-tail isomers. The mixture is converted into easily separable products upon treatment with concentrated hydrochloric acid at room temperature. Only di- and tri-substituted double bonds react with hydrochloric acid, the terminal monosubstituted double bonds remaining unaffected. While the head-to-head dimer

160 remains unchanged on treatment with hydrochloric acid, the tail-to-tail dimer *163* forms the dichloride *165* and the desired head-to-tail dimers *161* and *162* provide the monochloride, namely 7-chloro-3,7-dimethyl-1-octene (*164*) in 84% yield based on the head-to-tail dimers present (Scheme 46). The monochloride *164* is converted into α- and β-citronellols (*159, 166*). Linalool (*167*) is prepared from *164:*

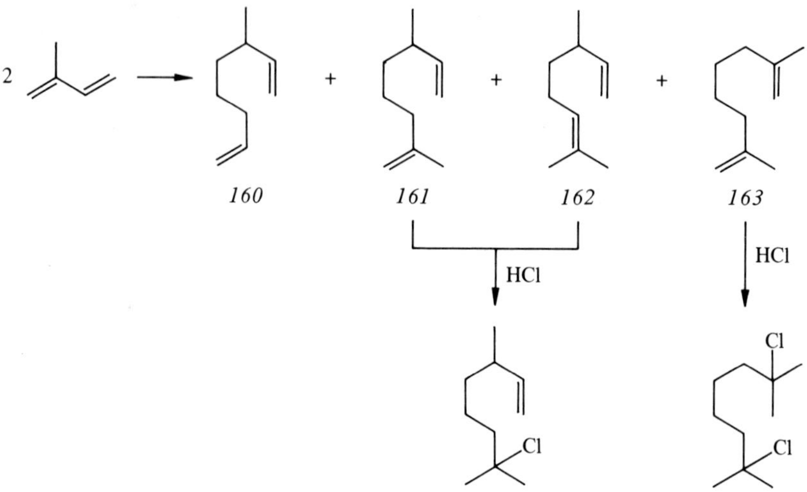

160 161 162 163

Scheme 46 164 165

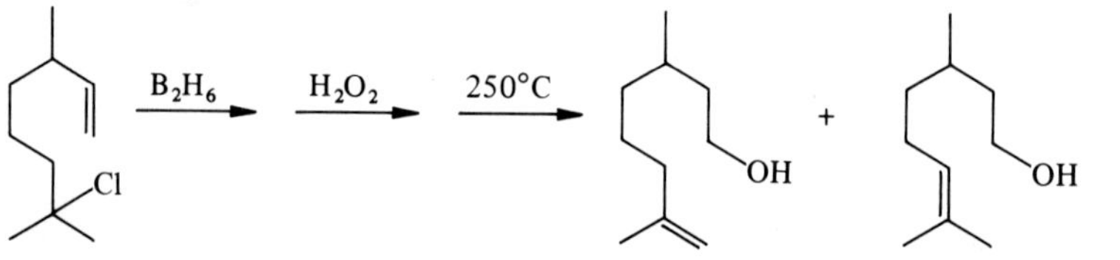

164 166 159

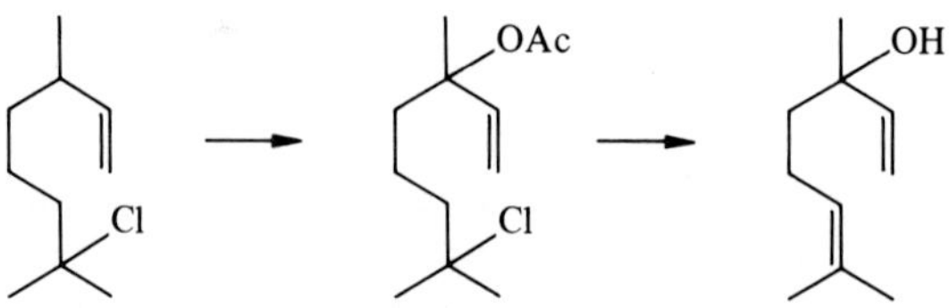

164 167

Reaction of 2,3-dimethyl-1,3-butadiene with methyl acetoacetate is carried out with $PdCl_2$ as a catalyst in the presence of sodium phenoxide. When triphenylphosphine is employed, a 1 : 2 adduct is obtained. On the other hand, the use of 3-methyl-1-phenyl-Δ^3-phospholene (*168*) at 100 °C causes formation of a 1 : 1 adduct to give

3-methoxycarbonyl-5,6-dimethyl-5-hepten-2-one (*169*) from which 5,6-dimethyl-5-hepten-2-one (*170*) is formed (Scheme 47). This compound is a useful intermediate in the α-irone synthesis[64].

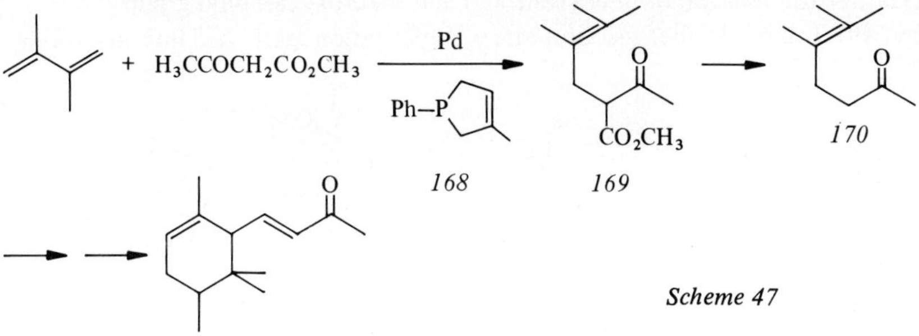

168 *169* *170*

Scheme 47

3.2 Reactions of Allylic and Alkenyl Esters and Ethers

Exchange reaction of allylic esters and ethers with nucleophiles is carried out by the catalytic action of $Pd[PPh_3]_4$ or $Pd(OAc)_2/PPh_3$[65,66]. The side chain of ecdysone

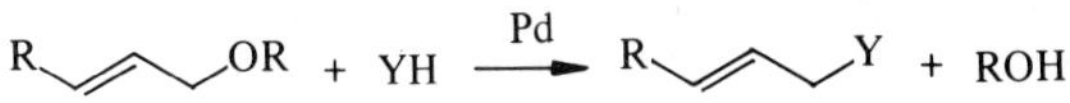

Y = nucleophile

(*172*) is formed by stereocontrolled displacement of methyl phenylsulfonylacetate at the allylic acetate moiety of steroid *171*; this displacement reaction, catalyzed by $Pd[PPh_3]_4$, representing a key step[67] (Scheme 48). It has been found that the

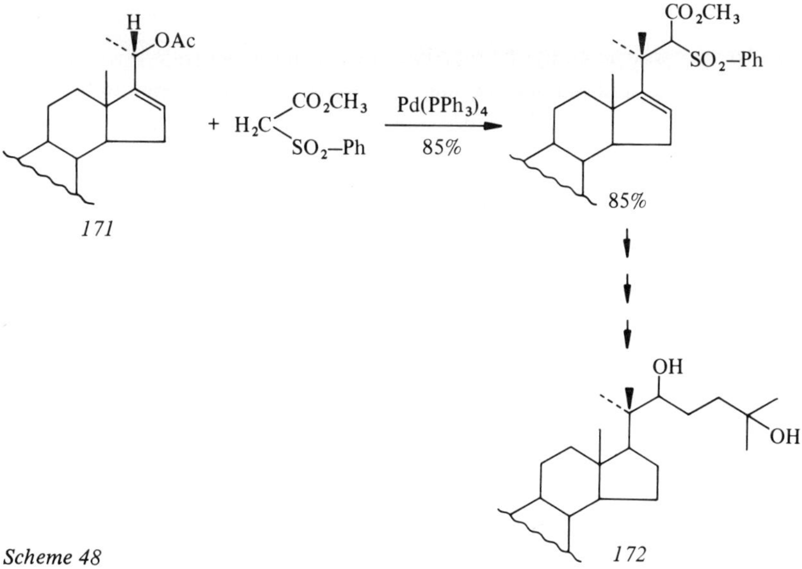

171

85%

172

Scheme 48

replacement of the acetoxy group by a carbon nucleophile proceeds with retention of configuration. Thus, the conversion of 5 α-androstane-3,17-dione (*173*) to 24,25-dehydro-18-nor-5α-cholestan-3-one (*174*) is carried out by reaction of the allylic acetate with methyl phenylsulfonylacetate catalyzed by a Pd^0 complex, followed by alkylation and removal of phenylsulfonyl and methoxycarbonyl group[68, 69] (Scheme 49). Intramolecular reaction offers a cyclization method. Thus, macrolide

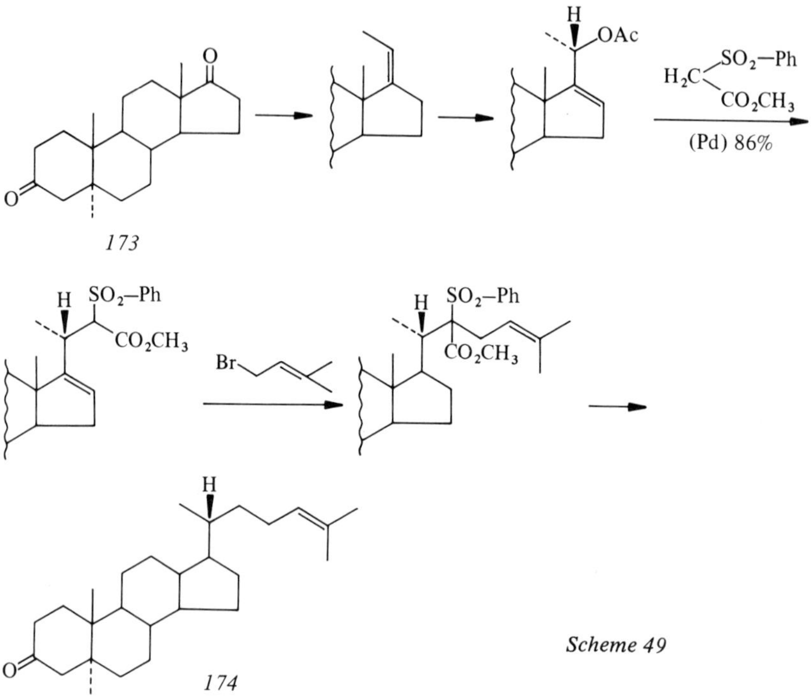

173

174

Scheme 49

skeletons *176* are constructed by the reaction of carbanions, generated from methyl phenylsulfonylacetate and sodium hydride, with the allylic acetate moiety in *175* using $Pd[PPh_3]_4$ as the catalyst[70] (Scheme 50). The cyclization method has been

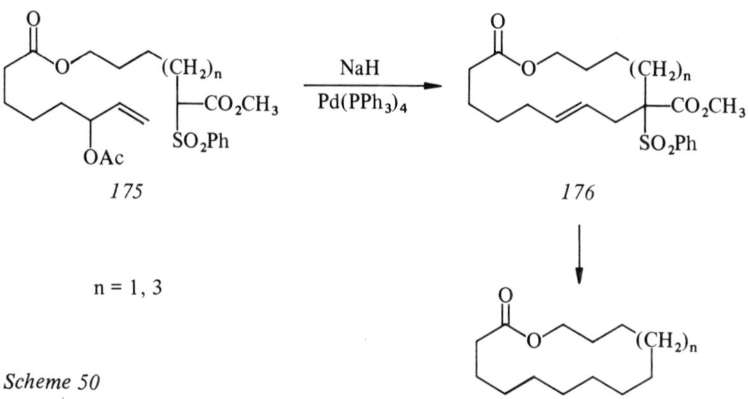

175

176

Scheme 50

applied to the synthesis of recifeiolide (*134*)[71] (Scheme 51). A THF solution of the anion generated from the precursor *177* is added slowly to a solution of $Pd(PPh_3)_4$ (9 mol%) at reflux temperature. The lactone *134* is obtained in 78% yield whereby the *E* isomer is produced stereoselectively and regioselectively without ten-membered lactone formation. The cyclization based on the intramolecular reaction

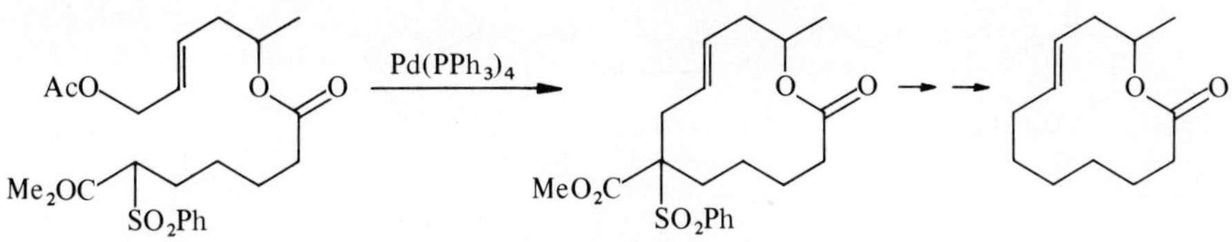

177 *134*

Scheme 51

of alkyl phenylsulfonylacetate with the allylic acetate moiety has been applied to the synthesis of ten-membered lactones (Phoracantholides). $Pd(PPh_3)_4$, coordinated by 1,2-bis-(diphenylphosphino)ethane, is used as the catalyst[72]. No eight-membered lactone is formed (Scheme 52). Similarly, eight- and nine-membered lactones *178*

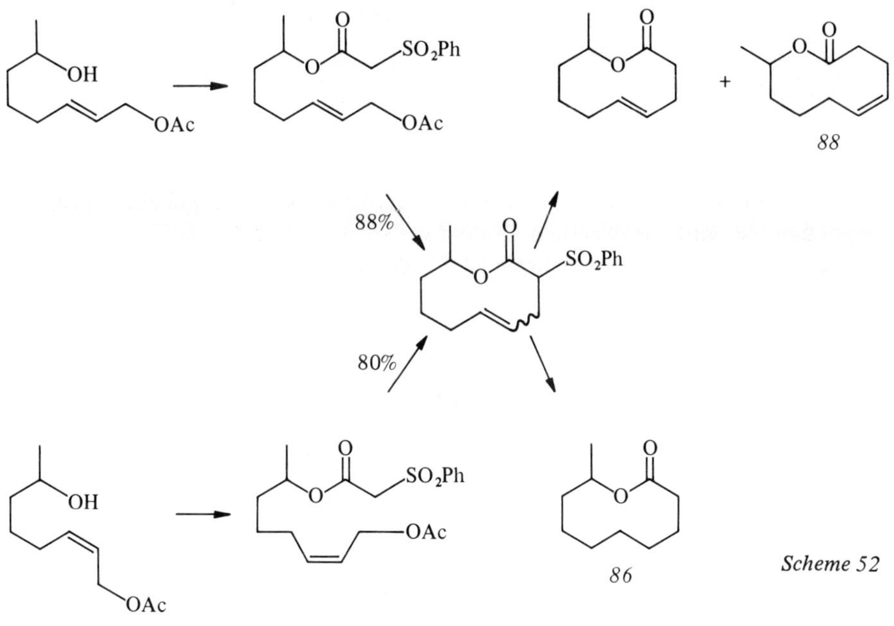

88

86

Scheme 52

are obtained without formation of six- and seven-membered lactones *179* (Scheme 53). Another application is the 11-membered ring formation of a humulene precursor *181*. The

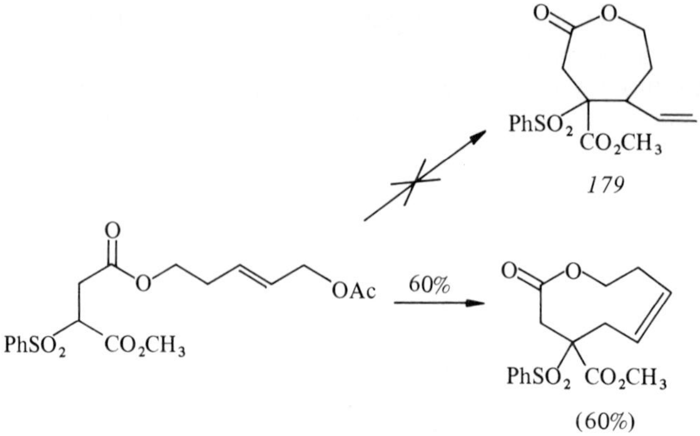

179

(60%)

Scheme 53

178

3-oxo ester moiety in *180* intramolecularly reacts with the allylic acetate residue by the catalysis of $Pd(PPh_3)_4$ (20 mol%) and 1,2-bis-(diphenylphosphino)ethane (20 mol%) to give the cyclized product *181* in 45% yield[73]:

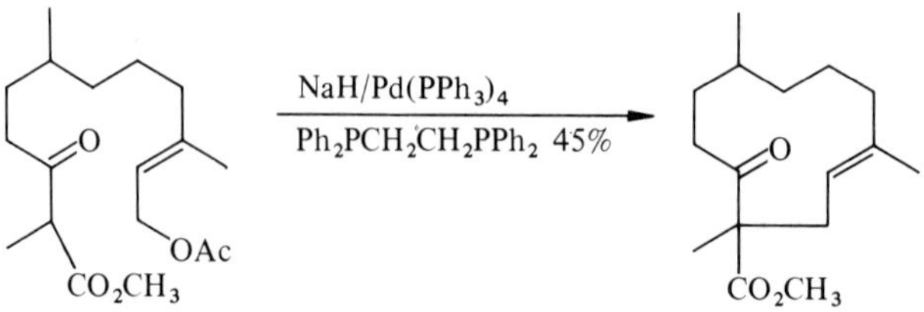

180

181

Intramolecular displacement reaction of allylic acetate with amines has been applied to the synthesis of certain alkaloid skeletons such as desethylibogamine and ibogamine (*183*) as shown in Scheme 54[74]. The reaction involves allylic rearrange-

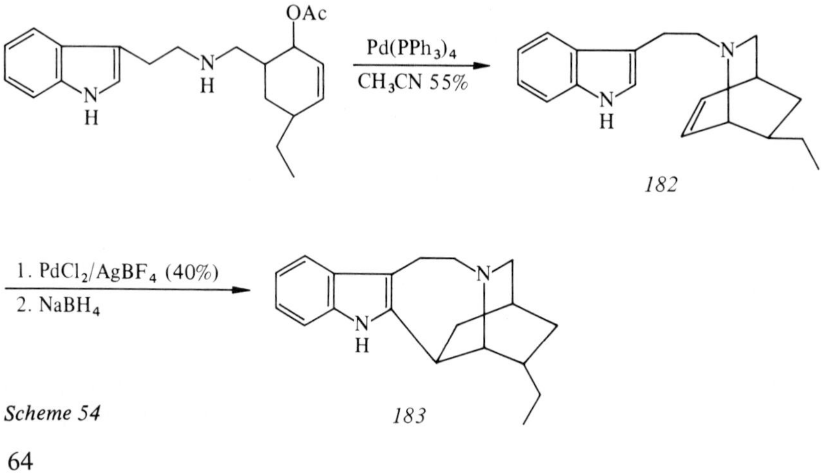

182

Scheme 54

183

ment and displacement. Reaction of olefins with benzene and other aromatic compounds yields styrene derivatives. Similarly the intramolecular reaction of the indole moiety with the double bond in *182* is promoted by $PdCl_2(CH_3CN)_2$ and $AgBF_4$ in acetonitrile in the presence of triethylamine. The intermediate palladium species is reduced with $NaBH_4$ to give the cyclized product *183* in 40% yield[75]. Various 1,3-diene epoxides are converted to dienols in high yields by heating with $Pd(PPh_3)_4$. Monoepoxides of simple cyclic 1,3-dienes possessing ordinary ring size afford β,γ-unsaturated ketones. By this method, an elegant synthesis of 4-hydroxy-2-cyclopentenone (*184*) has been achieved[76] (Scheme 55). This is an important intermediate in the prostaglandin synthesis. $Pd(PPh_3)_4$ is an extremely active catalyst for cyclopentadiene monoepoxide. Only 0.00013 mol% is sufficient to effect cleavage of the oxirane ring.

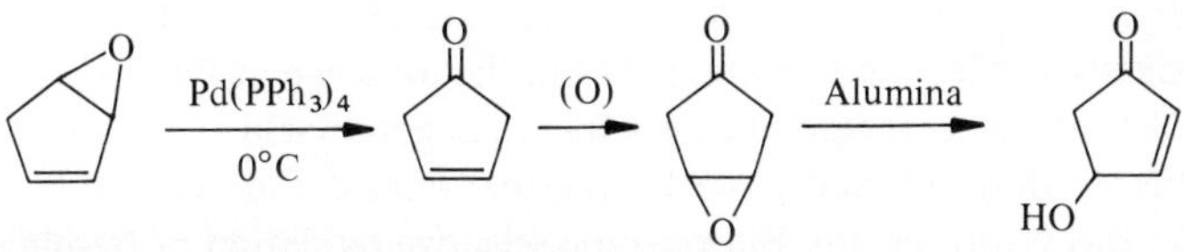

Scheme 55 *184*

The alcoholic component of vinyl ethers may be replaced by other alcohols in the presence of palladium salts.

$$CH_2{=}CHOR + R'OH \rightleftharpoons CH_2{=}CHOR' + ROH$$

$Pd(OAc)_2$ coordinated with 1,10-phenanthroline or 2,2'-bipyridine is a highly reactive catalyst for the exchange reaction of vinyl ethers. As shown in Scheme 56, only vinyl ethers are formed, the formation of acetals being completely suppressed[77]. Smooth exchange reaction under mild conditions has been applied to the total synthesis of rhizobitoxine (*185*). The key step is the palladium-catalyzed exchange reaction of the vinyl ether moiety[78].

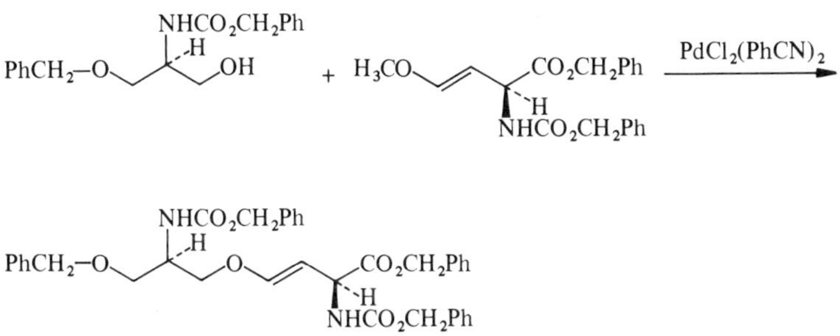

185 *Scheme 56*

3.3 Reactions of Organic Halides

Alkenyl, allyl, and aryl halides undergo oxidative addition to Pd^0 complexes to form alkenyl-, allyl-, and aryl-palladium σ-complexes which then react with carbon monoxide, alkenes and alkynes.

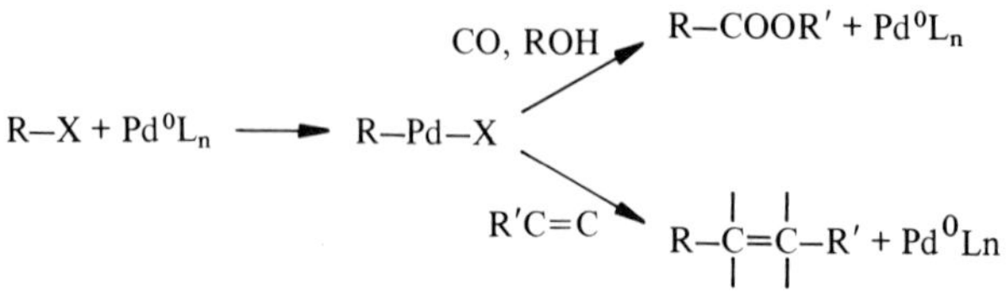

3.3.1 Allyl Halides

Allyl halides add to alkenylpentafluorosilicates (*186*) in the presence of $Pd(OAc)_2$ at room temperature to give the corresponding 1,4-dienes in good yield. This method has been applied to the synthesis of methyl(±)-11-hydroxy-trans-8-dodecenoate (*187*) (Scheme 57). In this synthesis, the *Wacker*-type selective oxidation of terminal

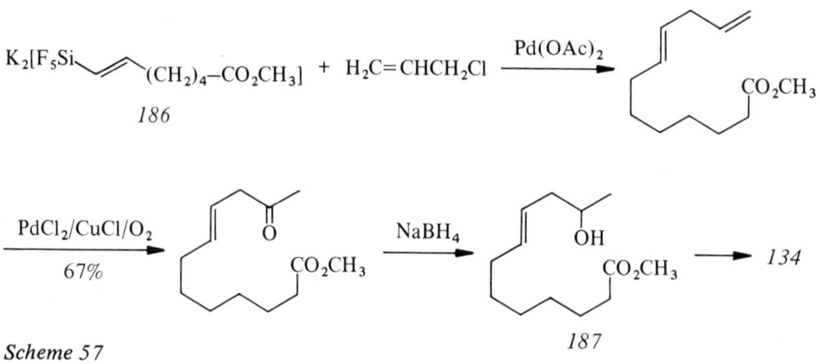

Scheme 57

olefinic double bonds is utilized[79]. The 1:1 addition reaction of allyl halides to various alkynes in the presence of $PdCl_2(PhCN)_2$ affords 1,4-pentadiene derivatives[80, 81]. For example, the reaction of 1-hexyne with allyl chloride gives 5-chloro-1,4-nonadiene in almost quantitative yield:

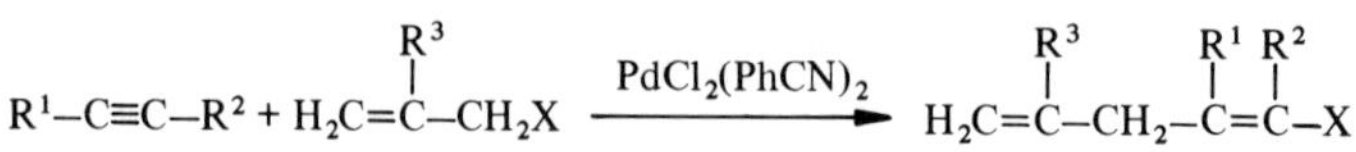

2,5-Undecanedione (*9*), a precursor of dihydrojasmone (*10*), is prepared by reaction of allyl chloride with 1-octyne followed by oxidation of the terminal double bond and hydrolysis of the resultant 2-oxo-4-undecen-5-yl-chloride with sulfuric acid[82] (Scheme 58).

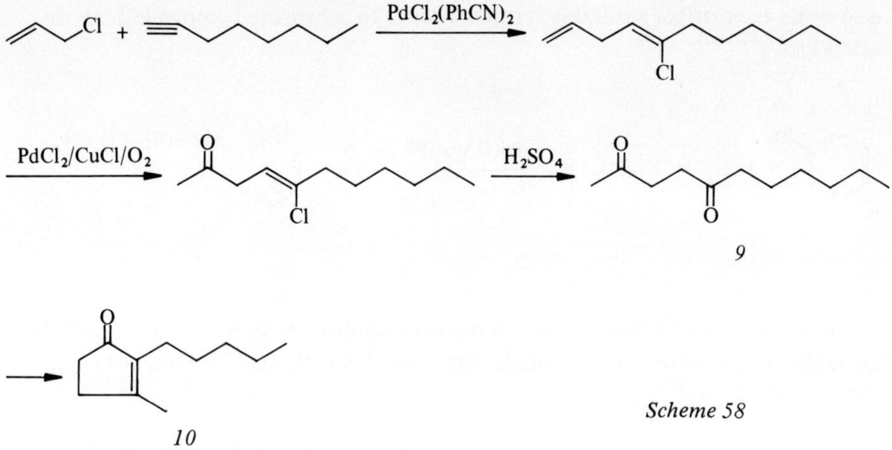

Scheme 58

3.3.2 Aryl and Alkenyl Halides

Aryl halides which are rather inert in usual organic reactions can undergo reactions by means of palladium catalysts. Thus, styrene and stilbene derivatives are obtained by reaction of olefins with aryl bromides at 125 °C using $Pd(OAc)_2$ (1 mol%) and tri-(*o*-tolyl)phosphine (2 mol%)[83]. The palladium-catalyzed vinylic substitution reaction is applicable to a variety of heterocyclic bromides including pyridine, thiophene, indole, furan, quinoline and isoquinoline[84]. Thus, reaction of 3-bromopyridine with 1-(3-butenyl)phthalimide at 100 °C gives 1-[4-(3′-pyridyl)-3-butenyl]-phthalimide (yield of mixed amine 57%, selectivity 68%) at 100 °C. This phthalimide is subsequently converted to nornicotine (*188*) (Scheme 59). The reaction of acrylic

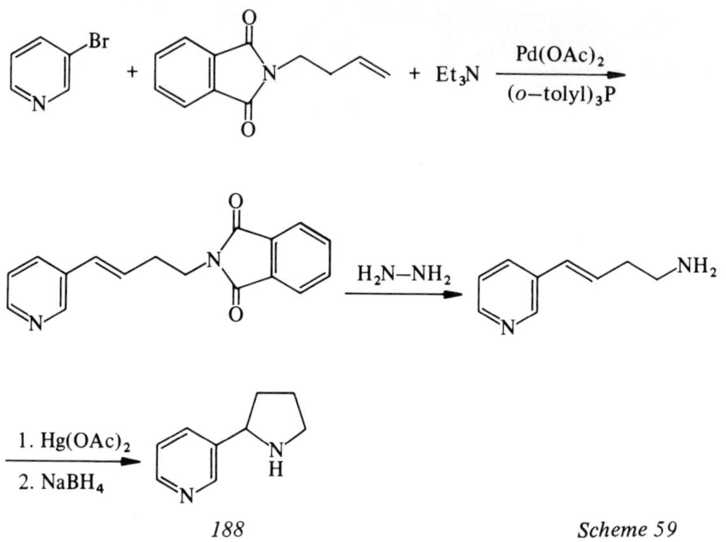

Scheme 59

acid or its ester offers a suitable synthetic route to substituted cinnamic acids or esters (*189*)[85, 86]:

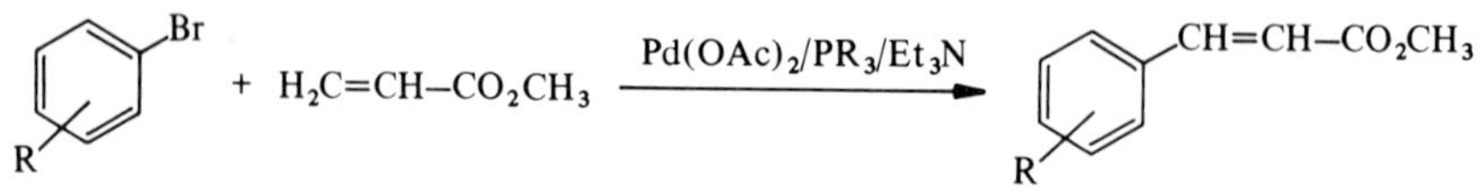

189

Intramolecular arylation reaction has been applied to the synthesis of indole and isoquinoline derivatives. For example, the 3-indolylacetic acid *190* has been synthesized as follows[87]:

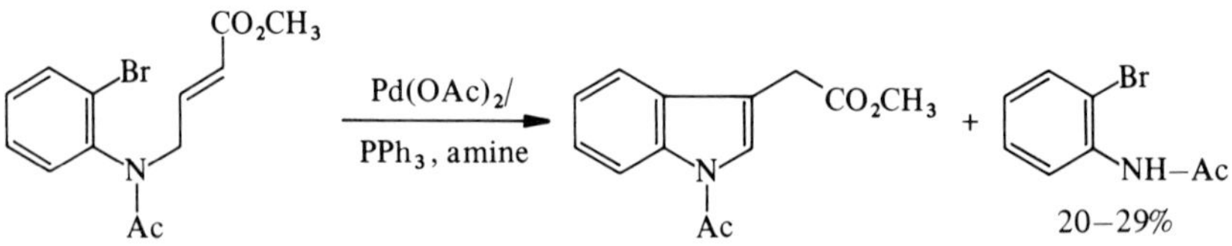

190
32–43%

20–29%

Reaction of bromobenzene with allylic alcohols by the catalysis of $Pd(OAc)_2$ and PPh_3 in the presence of tertiary amines affords hydrocinnamaldehydes (*191*)[88, 89] (Scheme 60). This reaction has been extended to 5-substituted 2- bromo-

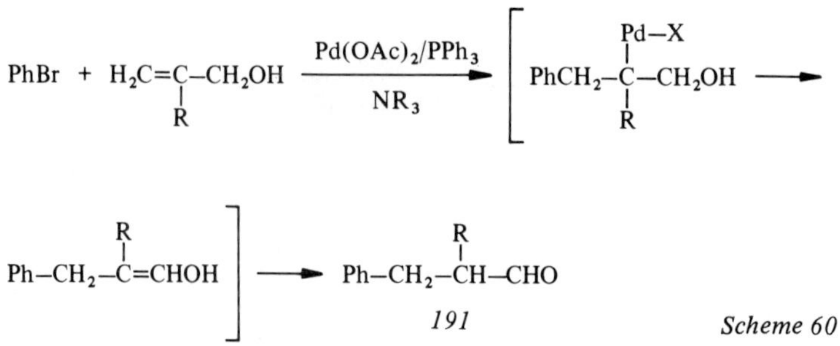

191

Scheme 60

thiophenes in order to prepare 9-oxo-2-decenoic acid (*4*)[90]. The addition product of the 2-bromothiophene and 3-hydroxy-1-butene (α-methallyl alcohol) is reductively cleaved with Raney-nickel and a double bond subsequently introduced into the conjugated position with respect to the carboxy group to give queen substance (*4*) (Scheme 61).

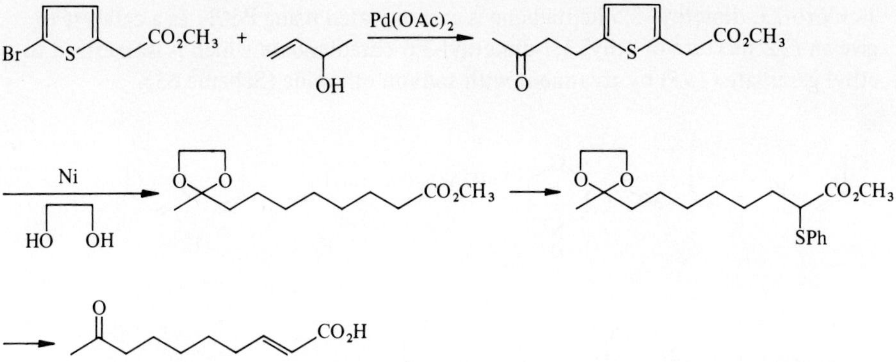

4

Scheme 61

Aryl bromide and iodides are carbonylated in the presence of alcohols to give esters in satisfactory yield. Carbonylation in the presence of primary amines at 1 atm affords arenecarboxamides in high yield[91]. This method has been extended to the

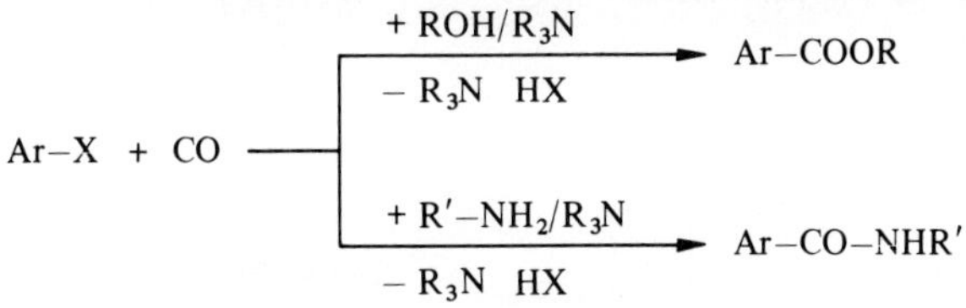

synthesis of sendaverine (*193*) (Scheme 62), a naturally occurring alkaloid. The key step is the carbonylation to give the 1-oxo-1,2-dihydroisoquinoline *192* (in 34.5% yield) which is reduced to sendaverine (*193*)[92]. The reaction of 1-bromo-2-methyl-

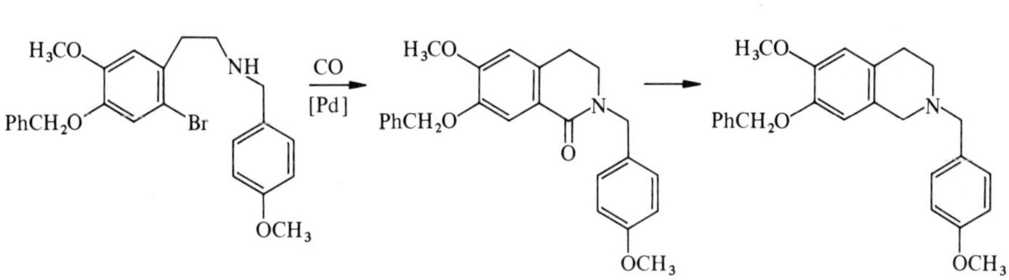

Scheme 62 *192* (35%) *193*

1-propene, isoprene and morpholine yields a mixture of adducts, the main product of which is the adduct *194*. This is converted to ethyl geraniate (*195*) by the following reaction sequence[93]. In the first step, the allylic 4-morpholinyl moiety in *194* is replaced by a chloro atom on treatment with methyl chloroformate. The resultant

1-chloro-2,6-dimethyl-2,5-heptadiene is carbonylated using $PdCl_2$ as a catalyst to give an *E/Z* mixture of ethyl 3,7-dimethyl-3,6-octadienoate which is isomerized to ethyl geraniate (*195*) by treatment with sodium ethoxide (Scheme 63).

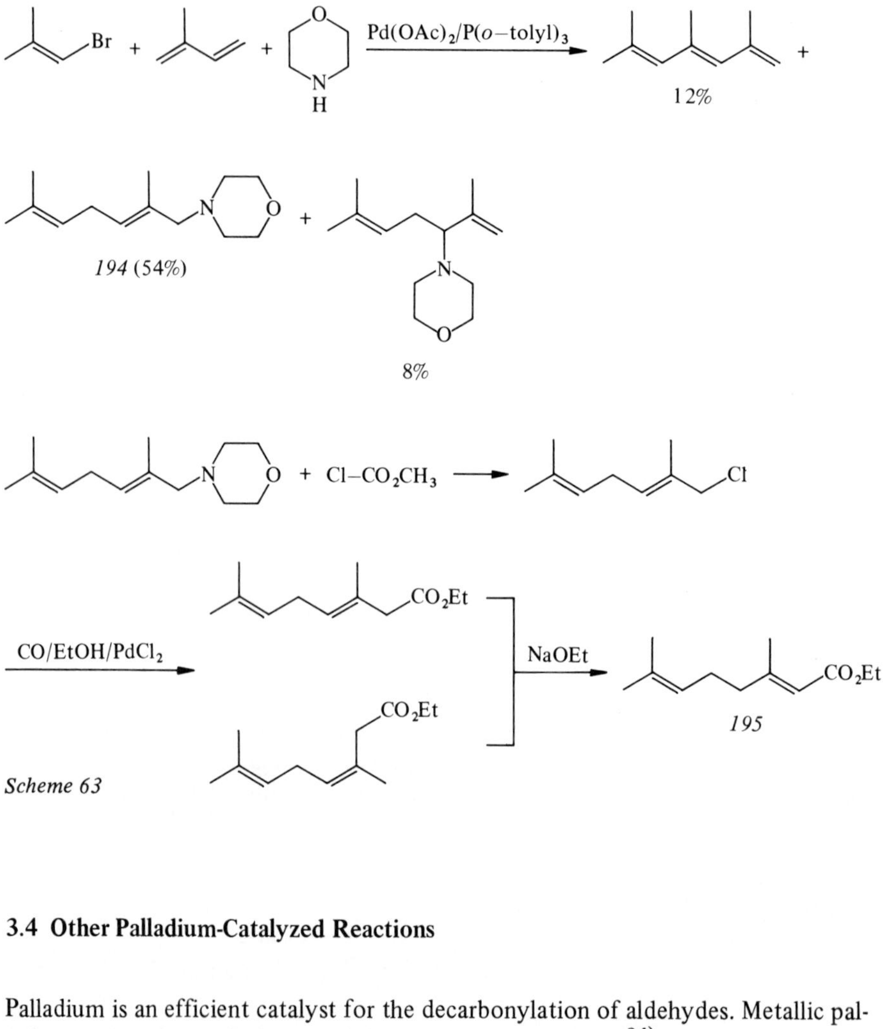

Scheme 63

3.4 Other Palladium-Catalyzed Reactions

Palladium is an efficient catalyst for the decarbonylation of aldehydes. Metallic palladium rather than palladium complexes is the active species[94].

$$R{-}CH_2{-}CH_2{-}CHO \xrightarrow[-CO,\ -H_2]{Pd} R{-}CH{=}CH_2 + R{-}CH_2{-}CH_3$$

Since the decarbonylation of aldehydes proceeds smoothly in high yield, it has been utilized for synthetic purposes. The first step of the five-step irone synthesis from α-pinene (*196*) involves formation of a formyl group by ozonolytic ring cleavage at the olefinic double bond. The resultant cis-pinonic aldehyde (*197*) is decarbonylated by heating with a palladium catalyst at 220 °C to give pinonone (*198*) and pinone-

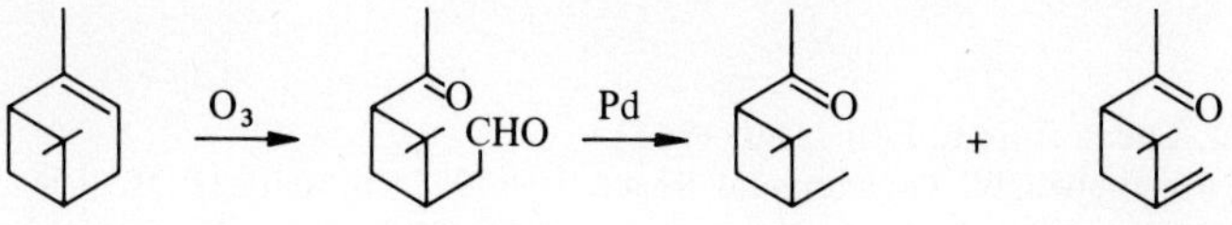

196 197 198 199

none (*199*) in 80% yield[95, 96]. Another application is the two-step preparation of apopinene from the easily available α-pinene[97]. The methyl group of α-pinene (*196*) is oxidized with selenium dioxide to a formyl group to form myrtenal (*200*) which is decarbonylated with palladium on barium sulfate at 195 °C to give apopinene (*201*) in an overall yield of 55%.

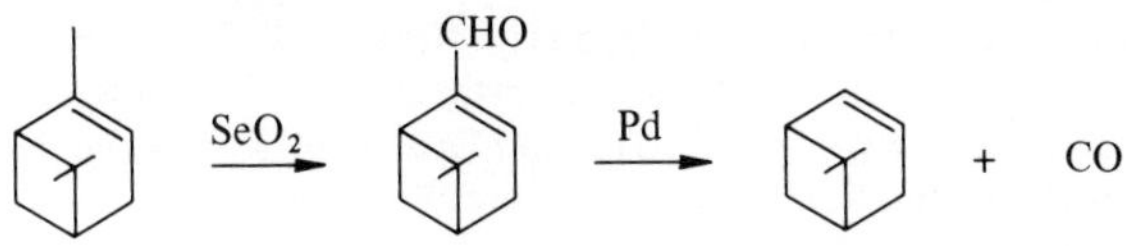

196 200 201

Depending on the catalytic species, palladium-catalyzed mono- and dicarbonylation of alkynes may be achieved. Monocarbonylation of acetylenic alcohols in the presence of thiourea is an elegant route to α-methylene-γ-butyrolactone *202*, the structure of which is widely distributed in certain natural products[98, 99]. The synthesis of a vernolepine derivative (*203*) has been attempted by this method[100]. Pro-

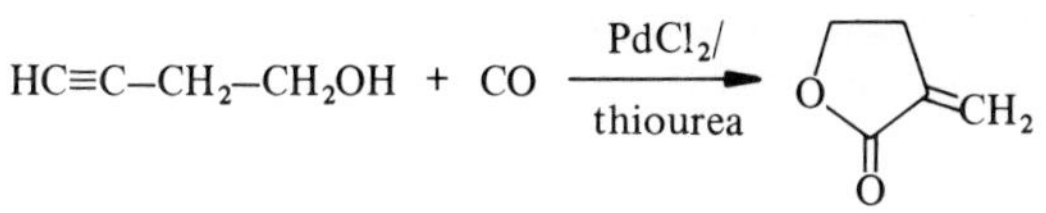

202

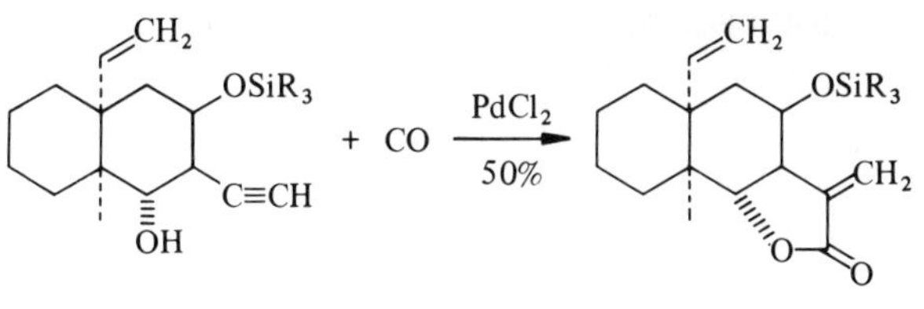

203

pargylic alcohols have been converted to the corresponding vinylic iodo-substituted alcohols which are carbonylated to α-butenolides. For example, 4-iodopent-3-en-2-ol (*204*) is carbonylated at 35 °C to 2(5*H*)-3,5-dimethylfuranone (*205*) by using $PdCl_2/(PPh_3)_2$ as a catalyst in the presence of potassium carbonate[101].

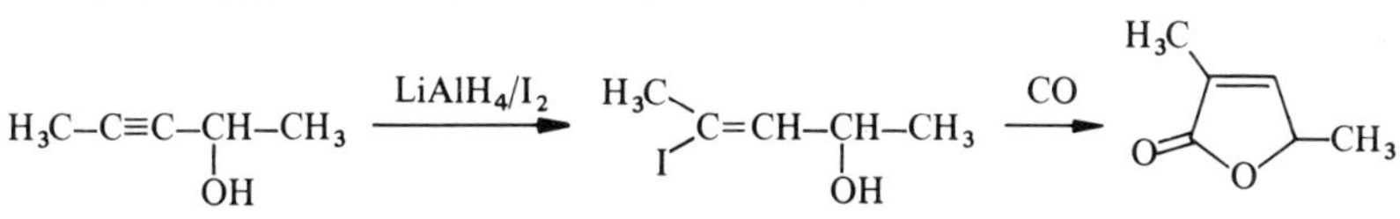

204 205

4 References

1. Smidt, J. et al.: Angew. Chem. (Intern. Ed.) *1*, 80 (1962)
2. Reviews by a. Jira, R., Freiesleben, W.: Organometal. React. *3*, 5 (1972); b. Maitlis, P. M.: The organic chemistry of palladium. New York: Academic Press 1971; c. Rylander, P. N. : Organic synthesis with noble metal catalysts. New York: Academic Press 1973; d. Tsuji, J.: Adv. Org. Chem. *6*, 109 (1969); e. Tsuji, J.: Organic synthesis by means of transition metal complexes. Berlin: Springer 1975; f. Tsuji, J.: Acc. Chem. Res. *2*, 144 (1969); g. Tsuji, J.: Acc. Chem. Res. *6*, 8 (1973); h. Tsuji, J.: Adv. Organometal. Chem. *17*, 141 (1979); i. Hüttel, R.: Synthesis *1970*, 225; j. Heck, R. F.: Organotransition metal chemistry. New York: Academic Press, 1974; k. Heck, R. F.: Topics Curr. Chem. *16*, 221 (1971); l. Heck, R. F.: Adv. Catalysis *26*, 323 (1977); m. Heck, R. F.: Acc. Chem. Res. *12*, 146 (1979); n. Henry, P. M.: Adv. Organometal. Chem. *13*, 363 (1975); o. Henry, P. M.: Acc. Chem. Res. *6*, 16 (1967); p. Aguilo, A.: Adv. Organometal. Chem. *5*, 321 (1967); q. Stern, E. W.: Catal. Rev. *1*, 73 (1967); r. Kozikowski, A. P., Wetter, H. F.: Synthesis *1976*, 561; s. Hegedus, L. S., in: New application of organometallic reagents in organic synthesis. Seyferth, D. (ed.). Amsterdam: Elsevier 1976; t. Trost, B. M.: Tetrahedron Report, *32*, 2615 (1977); u. Tsuji, J., Organic synthesis with palladium compounds. Berlin, Heidelberg, New York: Springer 1980
3. Clement, W. H., Selwitz, C. M.: J. Org. Chem. *29*, 241 (1964)
4. Lloyd, W. G., Luberoff, B. J.: J. Org. Chem. *34*, 3949 (1969)
5. Subramanian, C. S. et al.: Synthesis *1978*, 468
6. Hase, T. A., McCoy, K.: Synth. Commun. *1979*, 63
7. Tsuji, J., Masaoka, M., Takahashi, T.: Tetrahedron Lett. *1977*, 2267
8. Tsuji, J., Shimizu, I., Yamamoto, K.: Tetrahedron Lett. *1976*, 2975
9. Byrom, N. T., Grigg, R., Kongkathip, B.: Chem. Commun. *1976*, 216
10. Kasahara, A., Izumi, T., Ooshima, M.: Bull. Chem. Soc. Japan *47*, 2526 (1976)
11. Chapman, O. L. et al.: J. Am. Chem. Soc. *93*, 6696 (1971)
12. Kasahara, A., Fukuda, N.: Chem. Ind. *1976*, 485
13. Tsuji, J., Takahashi, H., Morikawa, M.: Tetrahedron Lett. *1965*, 4387
14. Trost, B. M. et al.: J. Am. Chem. Soc. *100*, 3407 (1978)
15. Trost, B. M., Dietsche, T. J., Fullerton, T. J.: J. Org. Chem. *39*, 737 (1974)
16. Trost, B. M. et al.: J. Am. Chem. Soc. *100*, 3426 (1978)
17. Manchand, P. S., Wong, H. S., Blount, J. F.: J. Org. Chem. *43*, 4769 (1978)
18. Trost, B. M., Verhoeven, T. R.: J. Am. Chem. Soc. *100*, 3435 (1978)
19. Heck, R. F.: J. Am. Chem. Soc. *90*, 5518, 5546 (1968)
20. Saito, R., Izumi, T., Kasahara, A.: Bull. Chem. Soc. Japan *46*, 1776 (1973)
21. Horino, H., Inoue, N.: Chem. Commun. *1976*, 500
22. Larock, R. C., Riefling, B., Fellows, C. A.: J. Org. Chem. *43*, 131 (1978)
23. Holton, R. A., Kojonaas, R. A.: J. Am. Chem. Soc. *99*, 4177 (1977)
24. Holton, R. A.: J. Am. Chem. Soc. *99*, 8083 (1977)
25. Takahashi, S., Shibano, T., Hagihara, N.: Tetrahedron Lett. *1967*, 2451
26. Walker, W. E. et al.: Tetrahedron Lett *1970*, 3817
27. Tsuji, J., Shimizu, I., Suzuki, H., Naito, Y.: J. Am. Chem. Soc. *101*, 5070 (1979)
28. Tsuji, J., Kobayashi, Y., Takahashi, T.: Tetrahedron Lett. *1980*, 483
29. Tsuji, J., Mandai, T.: Tetrahedron Lett. *1978*, 1817
30. Maruoka, K. et al.: J. Am. Chem. Soc. *99*, 7705 (1977)
31. Takahashi, T. et al.: J. Am. Chem. Soc. *100*, 7424 (1978)
32. Moore, B. P., Brown, W. V.: Aust. J. Chem. *29*, 1365 (1976)
33. Wakamatsu, T., Akasaka, K., Ban, Y.: Tetrahedron Lett. *1977*, 2755
34. Petrzilka, M.: Helv. Chim. Acta *61*, 3075 (1978)
35. Takahashi, T., Kasuga, K., Tsuji, J.: Tetrahedron Lett. *1978*, 4917
36. Takahashi, T. et al.: J. Am. Chem. Soc. *101*, 5072 (1979)
37. Tsuji, J., Tsuruoka, K., Yamamoto, K.: Bull. Chem. Soc. Japan *49*, 1701 (1976)
38. Tsuji, J., Mandai, T.: Chem. Lett. *1977*, 975
39. Tsuji, J., Yasuda, H., Mandai, T.: J. Org. Chem. *43*, 3606 (1978)

40. Tsuji, J., Kobayashi, Y., Shimizu, I.: Tetrahedron Lett. *1979*, 39
41. Reuter, J. M., Salomon, R. G.: J. Org. Chem. *42*, 3360 (1977)
42. Stoll, M., Rouve, A.: Helv. Chim. Acta *30*, 2019 (1947)
43. Tsuji, J. et al.: Chem. Lett. *1976*, 773
44. Tsuji, J., Kaito, M., Takahashi, T.: Bull. Chem. Soc. Japan *51*, 547 (1978)
45. Smutny, E. J.: J. Am. Chem. Soc. *89*, 6793 (1967)
46. Tsuji, J., Yamamkawa, T., Kaito, M., Mandai, T.: Tetrahedron Lett. *1978*, 2075
47. Mandai, T. et al.: Tetrahedron *35*, 309 (1979)
48. Tsuji, J., Yamakawa, T.: Tetrahedron Lett. *1979*, 3741
49. Mitsuyasu, T., Tsuji, J.: Tetrahedron *30*, 831 (1974)
50. Tsuji, J., Yamakawa, T., Mandai, T.: Tetrahedron Lett. *1978*, 565
51. Tsuji, J., Mandai, T.: Tetrahedron Lett. *1977*, 3285
52. Tsuji, J., Mori, Y., Hara, M.: Tetrahedron *28*, 3721 (1972)
53. Billups, W. E., Walker, W. E., Schields, T. C.: Chem. Commun. *1971*, 1067
54. Tsuji, J., Yasuda, H.: J. Organometal. Chem. *131*, 133 (1977)
55. Tsuji, J. et al.: Bull. Chem. Soc. Japan *51*, 1915 (1978)
56. Tsuji, J. et al.: Tetrahedron Lett. *1979*, 2257
57. Hata, G., Takahashi, K., Miyake, A.: J. Org. Chem. *36*, 2116 (1971)
58. Tsuji, J. et al.: Tetrahedron Lett. *1977*, 1917
59. Tsuji, J. et al.: Bull. Chem. Soc. Japan. *50*, 2507 (1977)
60. Tsuji, J., Kasuga, K., Takahashi, T.: Bull. Chem. Soc. Japan *52*, 216 (1979)
61. Yamazaki, H.: Abst. Symposium Organometal. Chem. *1971*, 40
62. Hidai, M. et al.: Chem. Commun. *1975*, 170; Synthesis *1977*, 334
63. Neilan, J. P. et al.: J. Org. Chem. *41*, 3455 (1976)
64. Watanabe, S., Suga, K., Fujita, T.: Can. J. Chem. *51*, 848 (1973)
65. Takahashi, K., Hata, G., Miyake, A.: Bull. Chem. Soc. Japan *46*, 1012 (1973)
66. Atkins, K. E., Walkers, W. E., Manyik, R. M.: Tetrahedron Lett. *1970*, 3821
67. Trost, B. M., Matsumura, T.: J. Org. Chem. *42*, 2036 (1977)
68. Trost, B. M., Verhoeven, T. R.: J. Am. Chem. Soc. *98*, 630 (1976)
69. Trost, B. M., Verhoeven, T. R.: J. Am. Chem. Soc. *100*, 3435 (1978)
70. Trost, B. M., Verhoeven, T. R.: J. Am. Chem. Soc. *99*, 3867 (1977)
71. Trost, B. M., Verhoeven, T. R.: Tetrahedron Lett. *1978*, 2275
72. Trost, B. M., Verhoeven, T. R.: J. Am. Chem. Soc. *101*, 1595 (1979)
73. Kitagawa, Y. et al.: J. Am. Chem. Soc. *99*, 3865 (1977)
74. Trost, B. M., Genet, J. P.: J. Am. Chem. Soc. *98*, 8516 (1976)
75. Trost, B. M., Godleski, S. A., Genet, J. P.: J. Am. Chem. Soc. *100*, 3930 (1978)
76. Suzuki, M., Oda, Y., Noyori, R.: J. Am. Chem. Soc. *101*, 1623 (1979)
77. McKeon, J. E., Fitton, P., Griswold, A. A.: Tetrahedron *28*, 227, 233 (1972)
78. Keith, D. D., Tortora, J. A., Ineichen, K., Leimgruber, W.: Tetrahedron *31*, 2633 (1975)
79. Yoshida, J., Tamao, K., Takahashi, M., Kumada: Tetrahedron Lett. *1978*, 2161
80. Kaneda, K. et al.: Tetrahedron Lett. *1974*, 1067
81. Kaneda, K. et al.: J. Org. Chem. *44*, 55 (1979)
82. Tsuji, J., Yasuda, H.: Synth. Commun. *1978*, 103
83. Plevyak, J. E., Heck, R. F.: J. Org. Chem. *43*, 2454 (1978)
84. Franck, W. C., Kim, Y. C., Heck, R. F.: J. Org. Chem. *43*, 2947 (1978)
85. Mizoroki, T., Mori, K., Ozaki, A.: Bull. Chem. Soc. Japan *44*, 581 (1971)
86. Patel, B. A. et al.: J. Org. Chem. *42*, 3903 (1977)
87. Mori, M., Katsumi, K., Ban, Y.: Tetrahedron Lett. *1977*, 1037
88 Chalk, A. J., Magennis, S. A.: J. Org. Chem. *41*, 273, 1206 (1976)
89. Melpolder, J. B., Heck, R. F.: J. Org. Chem. *41*, 265 (1976)
90. Tamaru, Y., Yamada, Y., Yoshida, Z.: Tetrahedron Lett. *1978*, 919; Tetrahedron *35*, 329 (1979)
91. Schoenberg, A., Heck, R. F.: J. Org. Chem. *39*, 3327 (1974)
92. Mori, M., Chiba, K., Ban, Y.: Heterocycles *6*, 1841 (1977)
93. Patel, B. A., Heck, R. F.: J. Org. Chem. *43*, 3898 (1978)

94. Tsuji, J., Ohno, K.: J. Am. Chem. Soc. *90,* 94 (1968)
95. Eschinazi, H. E.: J. Am. Chem. Soc. *81,* 2905 (1959)
96. Conia, J. M., Faget, C.: Bull. Soc. Chim. France *1964,* 1963
97. Eschinazi, H. E., Pines, H.: J. Org. Chem. *24,* 1369 (1959)
98. Norton, J. R., Shenton, K. E., Schwartz, J.: Tetrahedron Lett. *1975*, 51
99. Murry, T. F., Varma, V., Norton, J. R.: Chem. Commun. *1976,* 907
100. Chavdarian, C. G. et al.: Tetrahedron Lett. *1976,* 1769
101. Cowee, A., Stille, J. K.: Tetrahedron Lett. *1979,* 133

Received October 10, 1979

Aflatoxin Chemistry and Syntheses

Paul Francis Schuda

Department of Chemistry, University of Maryland, College Park, MD 20742, U. S. A.

Table of Contents

A. Introduction to the Aflatoxins and Related Compounds

1 Introduction

a) The Mold Metabolites

The aflatoxins and their structurally similar relatives are members of a larger family of compounds called mycotoxins. As the name mycotoxin implies, these toxins are produced during the spoiling of foodstuffs, through the action of certain fungi. The molds *Aspergillus flavus, Aspergillus parasiticus,* and *Aspergillus versicolor* are primarily responsible for the presence of this class of compounds, and can occur on a very diverse sampling of edible items including peanuts, rice, wheat, corn, cottonseed, soybeans, cheese, cocoa beans, meats, wines, butter, and spices. This widespread distribution of materials suitable for fungus growth, coupled with the extremely high toxicity and carcinogenicity of many of these toxins produced has generated, and rightfully so, a considerable amount of chemical and biological interest and attention.

Perhaps the most widely known and well publicized outbreak of aflatoxicosis occured at the beginning of the last decade (1960) in England when a sizable number of young turkeys were afflicted with the previously unrecognized "Turkey-X Disease". It was later determined that these poultry deaths were caused by severe liver damage incurred[1–3] while the farm animals had been ingesting ground-nut meals containing rather significant and quite notable amounts of *Aspergillus flavus* mold[4–6]. Indeed, even more recently[7], aflatoxins have been ascertained to be present at detectable levels in several common commercial brands of peanut butter.

b) Toxicity and Carcinogenicity

From the preceding discussion, it becomes obvious that a considerable amount of economic importance is attached to the determination of the physiological effects of these mycotoxins. Additionally, safe limits for human ingestion would have to be set and enforced.

Being so widespread in nature, many investigations[8] into the aflatoxins brought to light the fact that they are acutely toxic and highly carcinogenic compounds[9–14], even when present in miniscule amounts.

Aflatoxin B_1 (*1*), perhaps the most widespread and well-known of the aflatoxins, has been shown to be highly toxic to a variety of animals to varying degrees[15, 16], cause chromosomal changes[17], and produce necrosis of the liver after only a single dose[18]. In addition to its high toxicity, it has been shown[14, 19–21] to be one of the most potent carcinogens known[22]. Several excellent reviews can be obtained that contain a myriad of references on the structural[23–28] and biological effects[8, 29–35, 11, 12] of mycotoxins, and a fine synopsis of the structural vs biological activity of the aflatoxins is also available[31].

c) Detection and Separation

The relative amount of the mycotoxins present in agricultural products is generally in the range of micrograms of toxin per kilogram of foodstuff. Therefore, one is usually dealing with miniscule amounts of materials which are distributed in a non-homogeneous fashion. The methods for the detection and separation of the aflatoxins generally rely on thin layer chromatographic analytical techniques, thereby taking advantage of the high level of fluorescence of the compounds. The experimental samples are compared with a series of standard references using authentic toxins or toxin containing samples. Many procedures for the separation and estimation of mycotoxins have been summarized by Jones[36)] in a review in which full experimental details and many references are given. Additionally, more recent investigations of analytical methods for the determination of aflatoxins[9–13, 37)] have resulted in the application of high pressure liquid chromatography (HPLC) techniques to this problem[38–40)].

This area of research assumes a great deal of importance when one recognizes the relative ease of formation and the widespread occurence of the *Aspergillus* molds, and the extremely high physiological activity, both toxic and carcinogenic, of the mycotoxins thereby produced.

d) Control and Regulation

The development of increased accuracy and precision in the quantitative detection of these compounds saturates the first quanta of control and regulation in that it allows for direct and non-negotiable proof of the presence and amount of aflatoxin(s) in a given sample. The remaining need to be met is the regulation of the amount of these metabolites in consumables. This is currently done by the Food and Drug Administration in the United States (FDA)[7)] and was set at 20 ppb in 1969. Under consideration now is a further proposed reduction to the 15 ppb limit in peanuts. The results of this lowering of the FDA limits on the aflatoxins would present two alternatives to producers and distributors of staple foods. The first would relegate the use of the mold afflicted foods and grains to only non-food purposes such as cattle fodder or oil production. The other possibility would entail the implementation of a prevention-cure system in which either,

1) carefully controlled conditions of temperature and humidity during storage and transportation of foodstuffs would prevent the formation of the *Aspergillus* molds or,

2) the use of pharmacologically safe and effective detoxification agents on infected sources to remove, or to render inert, the toxins present.

Inclusive in such methods of chemical inactivation are treatments such as light[41, 42)], microbial[43)], roasting[44)], chemical agents (benzoyl peroxide, sodium hypochlorite, sodium hydroxide)[45)], and even normal processing techniques, as for example, freeze or spray drying[46)]. More details and references may be found in any of several reviews[9–12, 29, 32, 47–49)] concerning the elements and implications of the control of aflatoxins.

In any event, the control of the amounts of mycotoxins reaching the public will most certainly place a larger burden of responsibility, both regulative and economic, on all parties concerned with the general welfare of the consumer; i.e. the producers, the distributors, the government and its agencies, and finally the consumer himself.

2 Isolation and Structural Elucidation[50)]

a) Aflatoxins B_1(*1*), B_2(*8*), G_1(*9*), and G_2(*10*)

The aflatoxins B_1(*1*), B_2(*8*), G_1(*9*), and G_2(*10*) occur as the major highly active constituents of the *Aspergillus flavus* species and have been isolated and separated by a number of workers[2, 5, 19, 32, 51–53)]. The next problem to be assaulted was the determination of the chemical structures of these metabolites. Contributions to this effort were made by several groups[19, 54–58)].

The results of Buchi and co-workers at M. I. T. afforded[54, 55)] the complete structures of the major components, aflatoxins B_1(*1*) and G_1(*9*). The substance corresponding to aflatoxin B_1(*1*) was determined to have the molecular formula $C_{17}H_{12}O_6$; molecular weight = 312; melting point 268–269 °C dec.; $\lambda_{max.}^{ethanol}$ 223, 265, 362 mμ (ε 25,600; 13,400; 21,800); and $\nu_{max.}^{CHCl_3}$ 1760(s), 1665(w), 1630, and 1600 cm^{-1}. In addition, it readily absorbed three moles of hydrogen during catalytic hydrogenation to afford a product, (*2*), that exhibited the following characteristics: molecular formula $C_{17}H_{16}O_5$; molecular weight = 300; melting point 272–274 °C; $\nu_{max.}^{CHCl_3}$ 1705, 1625, and 1600 cm^{-1}; and an ultraviolet spectrum [$\lambda_{max.}^{ethanol}$ 255, 264, and 332 mμ (ε 8,500; 9,200; 13,900)] which was related in shape to that of the synthetic bicyclic coumarin (*3*), and even more closely identical with that of the tricyclic coumarin (*4*). The data presented also indicates the presence of an olefin, as well a carbonyl group which must be somehow conjugated because of its ability to undergo facile hydrogenolysis. Analysis of the infrared and n. m. r. spectra next led to the conclusion that the carbonyl group was attached to a saturated cyclopentane system on the coumarin, and since it was necessarily conjugated, required that it be attached at either C-3 or C-5. In order to clarify the situation, Buchi synthesized the two isomeric systems, 5,7-dimethoxycyclopentenone[3,2-c]coumarin (*5*) and 5,7-dimethoxycyclopentenone[2,3-c]coumarin (*6*) (vide infra). Comparison of the spectral characteristics (ultraviolet and infrared) of (*5*) and (*6*) with those of aflatoxin B_1(*1*) showed that the data from (*5*) was significantly different [$\nu_{max.}^{CHCl_3}$ 1726 cm^{-1} (br); $\lambda_{max.}^{ethanol}$ 245, 268, 356 mμ (ε 13,200; 8,700; 9,000)], and the data from (*6*) very similar [$\nu_{max.}^{CHCl_3}$ 1759 (s), 1685 (w), 1614, 1594, 1550 cm^{-1}; $\lambda_{max.}^{ethanol}$ 215, 257, 355 mμ (22,200; 9,650; 26,800)]. Thus, the partical structure of aflatoxin B_1 (*1*) was firmly established as (*7*).

The remainder of the aflatoxin system was identified by n. m. r. methods. The presence of a three proton singlet at 4.02 ppm indicated an aromatic methoxy group. Further, the pattern 6.89 (d, J = 7 Hz, 1); 6.52 (t, J = 2.5 Hz, 1), 5.53 (t, J = 2.5 Hz, 1), 4.81 (d of t, J = 2.5 and 7 Hz, 1), pointed to a 2,3-dihydrofuran residue as being present. Additional n. m. r. spectral comparisons of aflatoxin B_1(*1*) with those of the known sterigmatocystin (*30*)[59, 60)] (vide infra) confirmed that the structure was

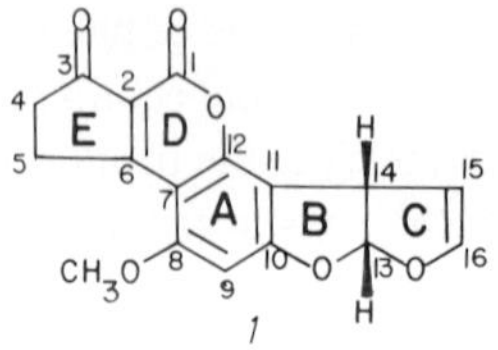

1

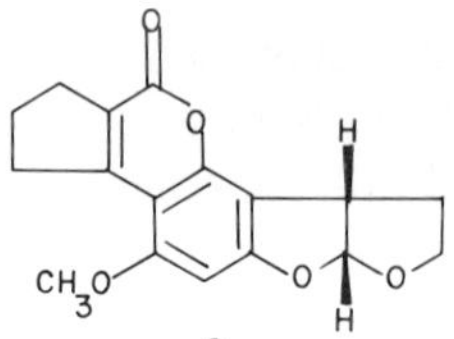

2

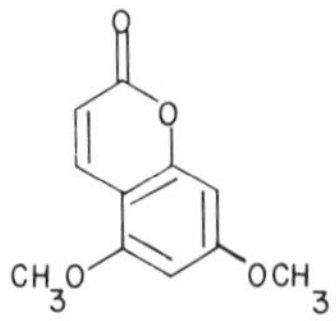

3

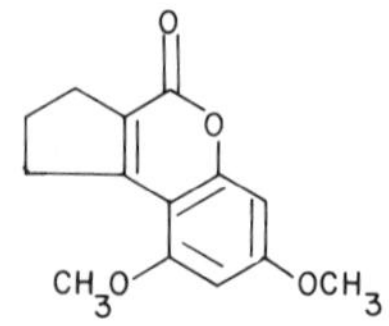

4

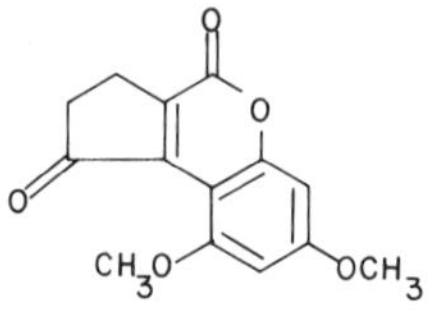

5

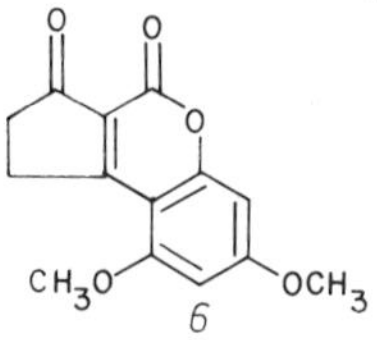

6

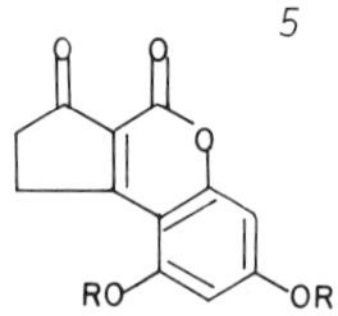

7

indeed that represented in (*1*). The aflatoxins B_2(*8*), G_1(*9*), and G_2(*10*) were also structurally elucidated in this study by comparative spectral means. Corroborative evidence of their structures was found in the experiments of other workers[56, 57)], and the structures of aflatoxins B_2(*8*) and G_1(*9*) were proven beyond all doubt with the disclosure of X-ray diffraction studies[58, 61)]. In addition, these X-ray studies showed that the junction of the bis-furan B-C ring system was indeed cis fused. Heretofore, it had only been assumed that this relative stereochemistry was cis, since it was known that a trans 5,5 ring system is not a very stable arrangement. Thus, the basic ring skeleton of the aflatoxin system was firmly established.

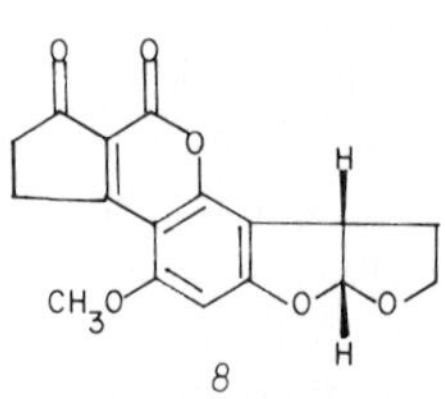

8

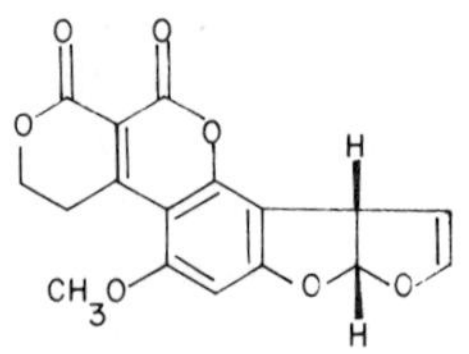

9

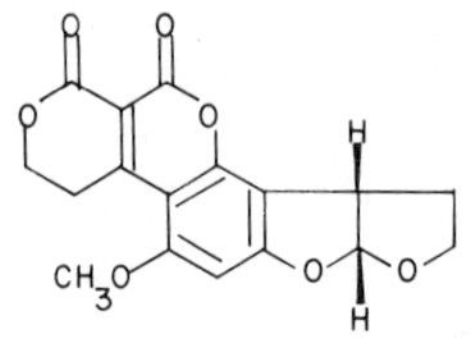

10

It remained now, only to ascertain the absolute stereochemistry of the two adjacent asymmetric centers in the furo-furan sector of the molecule. This obstacle was overcome by Buchi[62)] through the chemical degradation of aflatoxin B_1(*1*) to (+)-(S)-2-methylbutanoic acid, and subsequent transformation (+)-(S)-2-methylbutanoic amide, which was compared with an authentic sample of the latter compound. This truly elegant piece of degradative work thereby also gives the absolute configurations of the aflatoxins B_2(*8*), G_1(*9*), and G_2 (*10*) by virtue of chemical transformations and circular dichroism comparisons, i.e. since the CD spectrum of aflatoxin G_1(*9*) is superimposable with that of aflatoxin B_1 (*1*), they must necessarily have identical absolute configurations. Also, since aflatoxins B_1(*1*) and G_1(*9*) can be readily converted into aflatoxins B_2(*8*) and G_2(*10*) respectively by controlled catalytic hydrogenation[19, 56, 57, 63] this method also gives the absolute configurations of these compounds. The absolute configuration of these toxins is drawn in the correct enantiomeric form.

b) The Remaining Aflatoxins: B_{2a}(*11*), G_{2a}(*12*), M_1(*13*), M_2(*14*), GM_1(*15*), B_3(*16*), R_0(*17*), P_1(*18*), Q_1(*19*), RB_1(*20*), RB_2(*21*), and D_1(*22*)

1 Aflatoxins B_{2a}(11) and G_{2a}(12)

Aflatoxins B_{2a}(*11*) and G_{2a}(*12*) are the hemiacetals of aflatoxins B_1(*1*) and G_1(*9*) respectively and can be isolated as a metabolite from *Aspergillus flavus* molds[64], or, in the case of B_{2a}(*11*), as a liver metabolite of aflatoxin B_1(*1*)[65]. The structures were identified by spectral comparisons with compounds that were prepared by the acid catalyzed hydroxylation[66–68] and acetoxylation of aflatoxins B_1(*1*) and G_1(*9*) respectively[69, 70].

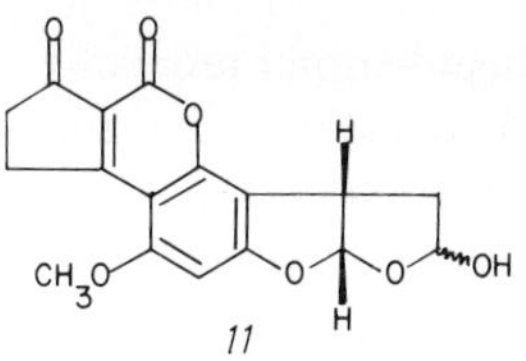

11

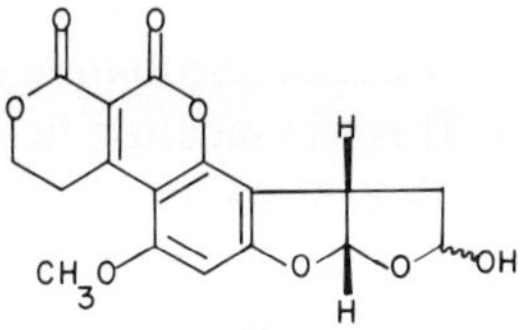

12

2 Aflatoxins M_1(13) and M_2(14)

It had been found by Allcroft[71] that certain isolated extracts of the milk produced by lactating cattle that had been fed sublethal levels of the aflatoxins, induced the formation of the liver lesions reminiscent of aflatoxicosis[72, 73]. The toxins were isolated and called the "milk toxins", which were later redesignated as the aflatoxins M[18, 74, 75]. Indeed, it was also demonstrated that the milk toxins were present in the original *Aspergillus flavus* mold as minor components[72].

The structures of the milk toxins were determined after their isolation from sheep urine[74, 76] and milk[72, 77]. The two constituents were separated, and labelled aflatoxin M_1(*13*) (less polar) and aflatoxin M_2(*14*) (more polar). The molecular formula of aflatoxin M_1(*13*) was determined as $C_{17}H_{12}O_7$; i.e. containing one more oxygen atom than aflatoxin B_1(*1*). The infrared spectrum ($\nu_{max.}^{CHCl_3}$ 3425, 1760, 1690 cm^{-1}) indicated the presence of a hydroxyl group, and this fact was substantiated both by acetylation with acetic anhydride/pyridine to produce a monoacetate (m/e = 370; $\nu_{max.}^{CHCl_3}$ 1760, 1740, 1692 cm^{-1}), and by mass spectral analysis (m/e = 310 = $M^+ - H_2O$). The similarity of the ultraviolet spectrum [$\lambda_{max.}^{ethanol}$ 226, 265, 357 mμ (ε 23,100; 11,600; 19,00)] of M_1(*13*) to that of B_1(*1*) implied an identical chromophoric system. Nuclear Magnetic Resonance analysis [δ d_6DMSO 3.98 (s, 3), 5.64 (d, J = 3 Hz, 1), 6.46 (s, 1), 6.78 (s, 1), 6.83 (d, J = 3 Hz, 1) ppm] showed that the vinyl ether protons were coupled only to each other, thereby indicat-

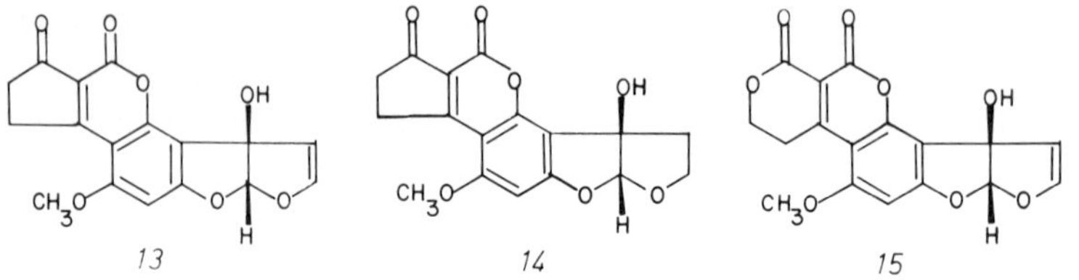

13 14 15

ing that the hydroxyl group was probably located in the junction of the furo-furan system at the benzylic position. Collectively, these results alluded to the structure of aflatoxin M_1 being (*13*). Aflatoxin M_2(*14*) was shown by spectroscopic means to be identical to dihydroaflatoxin M_1, obtained by partial catalytic hydrogenation of aflatoxin M_1(*13*) (Pd/C in acetic acid) ($\nu_{max.}^{CHCl_3}$ 3350, 1760, 1690 cm^{-1}; $\lambda_{max.}^{ethanol}$ 221, 264, 357 mμ (ε 20,000; 10,900; 21,000); m/e = 330). In addition, the nature of the tertiary hydroxyl group was confirmed by the failure of aflatoxin M_2(*14*) to undergo oxidation with chromium trioxide.

The implications of the seriousness of aflatoxicosis being caused by aflatoxin M_1(*13*) and/or aflatoxin M_2(*14*) are manifest when one considers that the widespread occurence of their metabolic precursors[73)] necessarily precludes the possibility of a proportionally equally widespread distribution of the perhaps harmful products in milk and milk products. This fact requires, therefore, both detection[9)], and control[11, 12)], since it has been determined that the acute toxicity of aflatoxins M_1(*13*) and M_2(*14*) is almost equal to that of the aflatoxins B_1(*1*) and B_2(*8*) respectively[23–26)].

3 Aflatoxin GM_1(15)

This metabolite was isolated by several groups[78–80)], and was assigned to be structure (*15*), the benzylic hydroxylated form of aflatoxin G_1(*9*). The structure was assigned on the basis of the similarity of spectra of the isolated compound with that of aflatoxin G_1(*9*), and the ready formation of a monoacetate under common acylation conditions.

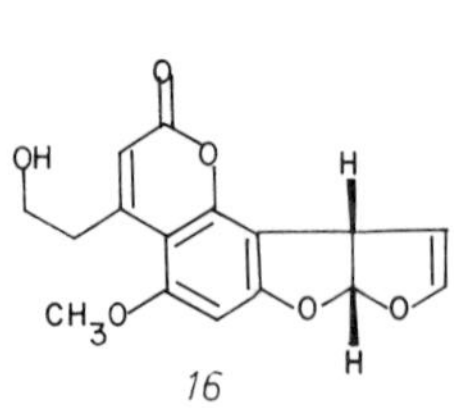

16

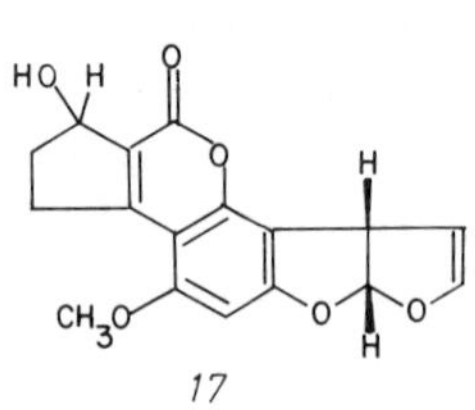

17

4 Aflatoxin B_3(16) (Parasiticol)

Parasiticol (*16*) was isolated by Heathcoate[80)] and Stubblefield[9)] from *Aspergillus flavus* cultures. Detailed n. m. r. analysis of this mycotoxin revealed it to be a close structural relative of aflatoxin G_1(*9*). The fact that the infrared spectrum indicated that the dilactone system of aflatoxin G_1(*9*) was absent, in conjuction with a wealth of additional chemical and spectral data, led to the conclusion that the structure of aflatoxin B_3 could be designated as (*16*). This structure could concievably be metabolically and/or chemically derived from aflatoxin G_1(*9*) by a hydrolysis-decarboxylation sequence.

5 Aflatoxin R_0(17) (Aflatoxicol)

Aflatoxicol occurs as one of the metabolic transformation products of aflatoxin B_1(*1*) via the reduction of the cyclopentanone carbonyl moiety[81]. The isolation and structure were done by Detroy[82], and the toxin identified as (*17*).

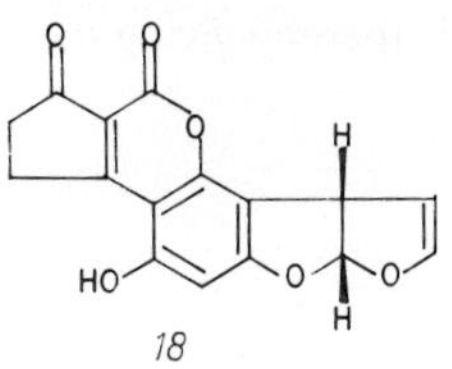

18

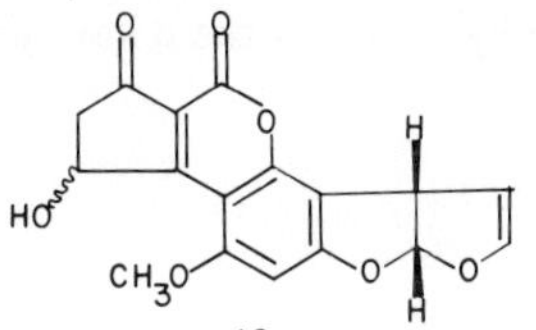

19

6 Aflatoxin P_1(18)

Yet another metabolite of aflatoxin B_1(*1*) is derived through the demethylation of the aromatic methyl ether. The resultant compound was isolated, and christened aflatoxin P_1[83, 84]. The structure was determined to be that depicted in (*18*).

7 Aflatoxin Q_1(19)

The in vitro metabolism of aflatoxin B_1(*1*) in the vervet monkey liver[85, 86] was demonstrated to produce a new compound. Spectral analysis, molecular composition data, and chemical tests showed the product to be an oxygenated derivative of aflatoxin B_1(*1*), which was distinctly different from aflatoxin M_1(*13*) by direct comparison. The data available suggested that the B_1(*1*) had been hydroxylated at the C-5 allylic position to yield 5-hydroxyaflatoxin B_1, or aflatoxin Q_1(*19*).

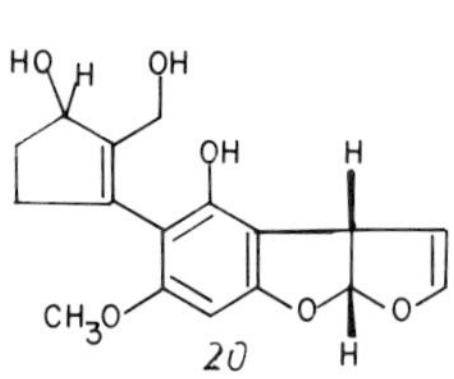

20

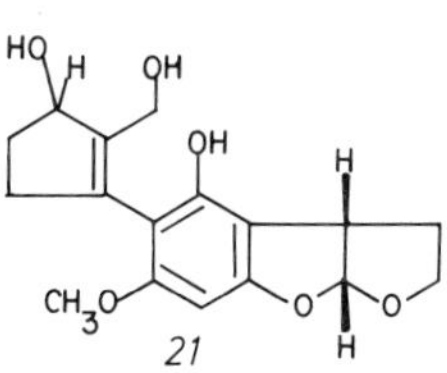

21

8 Aflatoxins RB_1(20) and RB_2(21)

The structures of the products resulting from the chemical reduction of aflatoxins B_1(*1*) and B_2(*8*), by a hydride source (sodium borohydride) are designated as aflatoxins RB_1(*20*) and RB_2(*21*) respectively. They are arrived at by opening of the coumarin lactone ring followed by the reduction of the two carbonyl systems[87]. Several reviews with accompanying references regarding the chemistry of the aflatoxin system are in the literature[31, 88].

9 Aflatoxin D_1(22)

The ammoniation of aflatoxin B_1(*1*) was studied as a means of possible detoxification of contaminated foodstuffs. Aflatoxin D_1 (*22*) arises through a mechanism whereby the coumarin lactone undergoes ammonolysis with ammonium hydroxide. The resulting ammonium salt is then postulated to decompose to the carboxylic acid, which decarboxylates by virtue of the β-keto group at C-3, thereby giving the desired structure (*22*)[89, 90].

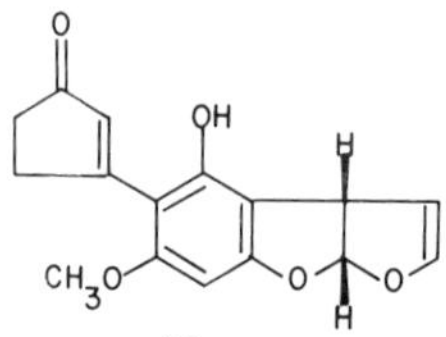

22

c. Related Mycotoxins

1 Sterigmatocystin (*30*)

A series of metabolites having structures that are closely related to the aflatoxins should now be examined. In point of fact, the structural studies and investigations that led to the aflatoxin B_1 (*1*) skeleton were illuminated considerably by comparison of the n. m. r. spectrum with that of the previously known, and structurally elucidated molecule called sterigmatocystin (*30*).

Sterigmatocystin (*30*) was isolated as a metabolite of *Aspergillus versicolor*[91–93] and other molds[94], and structurally postulated to be (*23*)[95, 96]. This structure was later found to be incorrect and an amended version put forth[59] (vide infra).

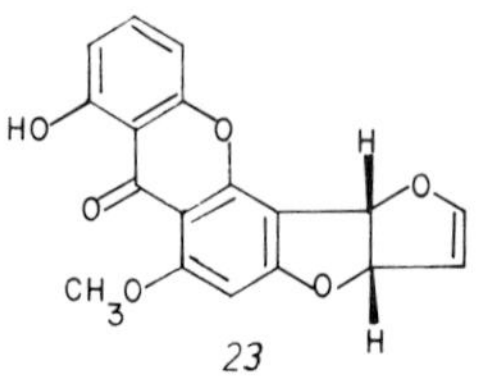

23

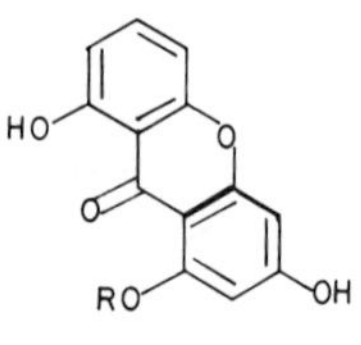

24 R = H
25 R = CH_3

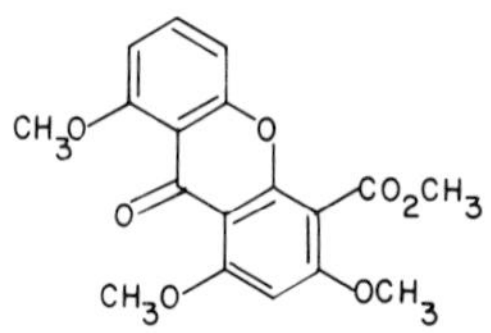

26

One sector of the sterigmatocystin (*30*) skeleton was brought to light by the chemical degradation[59, 91–98] of the spectroscopically determined xanthone[99] system; (1) Reaction of sterigmatocystin (*30*) with aluminum chloride afforded the polyhydric xanthone (*24*); (2) The oxidation of sterigmatocystin (*30*) with persulfate, and subsequent decarboxylation yielded the methoxy dihydric xanthone (*25*). The structure of this product was verified by comparison with a purely synthetic[95] sample of the same compound. This result allowed for the ready determination of which phenolic oxygen of sterigmatocystin (*30*) existed as the methyl ether; (3) Persulfate oxidation of O-methylsterigmatocystin (*31*), followed by exhaustive methylation gave the triether (*26*)[96]. Collectively, these experiments provided evidence for the regiochemical structure of the xanthone piece, as well as the points of attachment for the remaining sector of the molecule [cf. (*27*)].

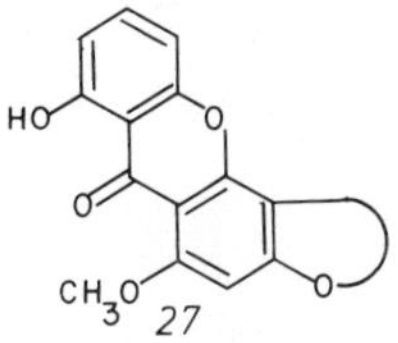
27

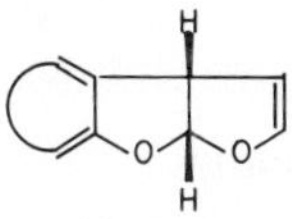
28

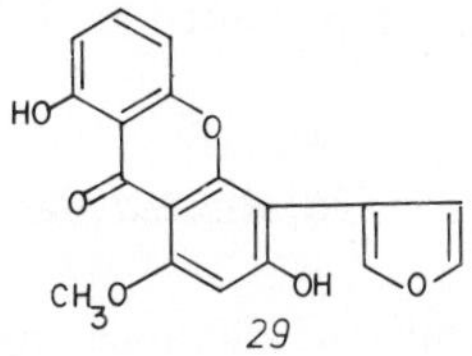
29

The second portion of sterigmatocystin (*30*) was now able to be assaulted. The presence of a vinyl ether was confirmed both spectroscopically, and chemically by the fact that one equivalent of hydrogen was absorbed when sterigmatocystin (*30*) was subjected to catalytic reduction. Sterigmatocystin is converted to an optically inactive compound labelled as isosterigmatocystin (*29*) when treated with sodium hydroxide. This isomeric compound (*29*) afforded a methyl ether which underwent Diels-Alder cycloaddition[59], thereby indicating the presence of a furan ring. Oxidative cleavage of isosterigmatocystin trimethyl ether gave (*26*) (after esterification), and two moles of formic acid, showing that the furan was 3-substituted. These results

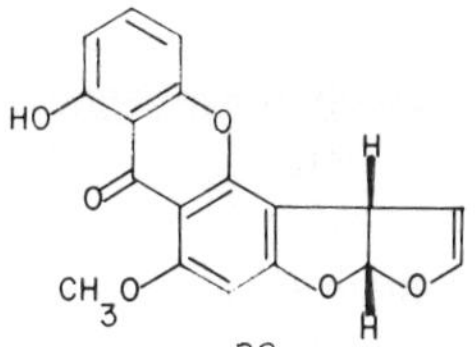
30

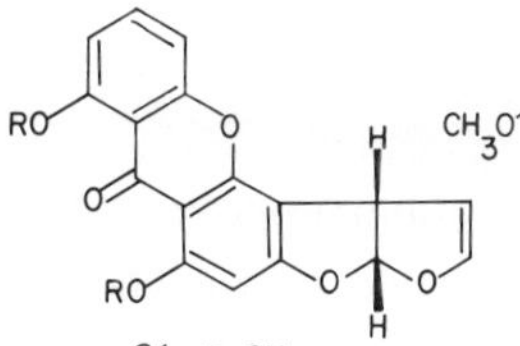
31 R=CH_3
32 R=H

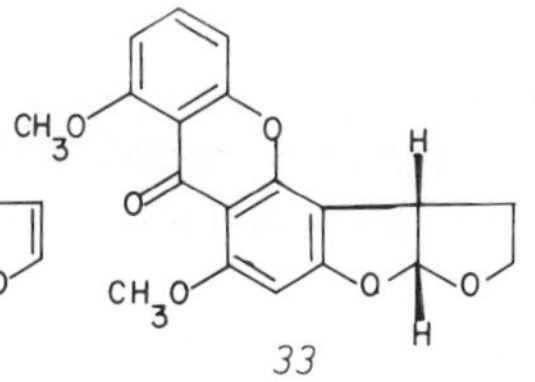
33

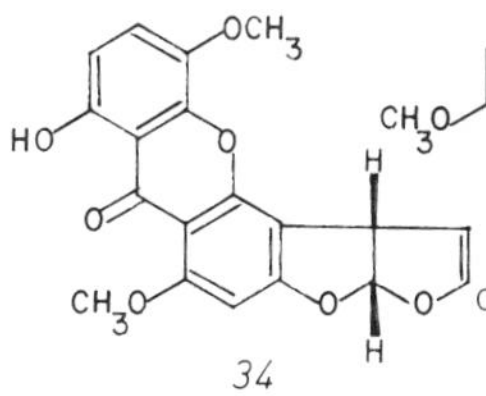
34

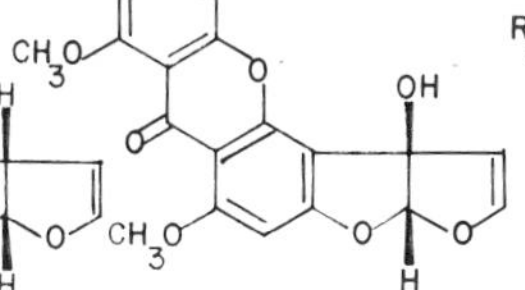
35

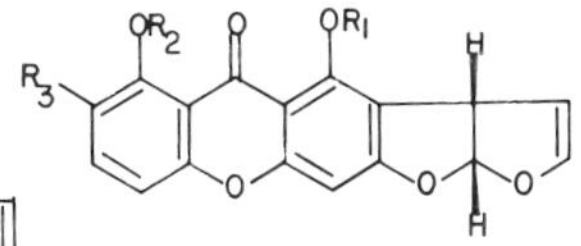
36 R's= CH_3's, Cl's, and H's

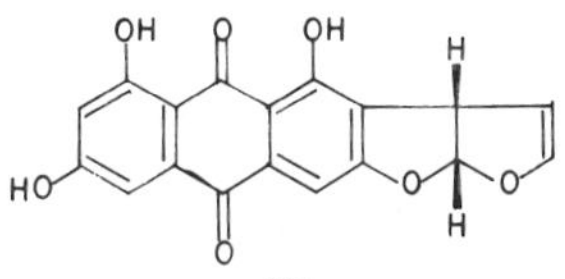
37

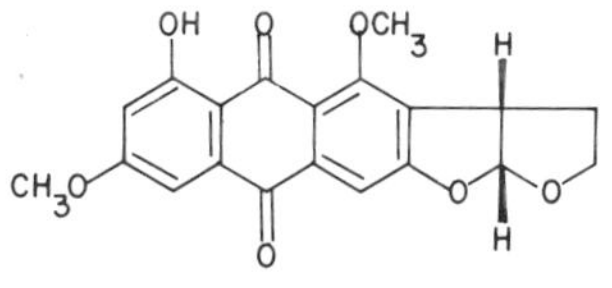
38

led to the formulation of the other partial structure section of sterigmatocystin (*30*), as (*28*). Additionally, the molecular arrangement of isosterigmatocystin was consequently deduced to be (*29*)[59].

Fusion of the two identified pieces, in the only manner feasible, finally allowed the representation of sterigmatocystin as (*30*)[59]. This structure was corroborated fully by the observed spectral characteristics, and was fully confirmed by a later X-ray diffraction study[100].

In order to ascertain the absolute configuration of sterigmatocystin (*30*), the methodology utilized by Buchi for the determination of the absolute configuration of aflatoxin B_1 (*1*)[62], was applied[101]. The final results showed the natural product to have the configuration represented in (*30*).

2. Other Sterigmatocystin (30) Related Metabolites

There are a notable number of additional mycotoxins which are structurally related to sterigmatocystin (*30*). The structural details were generally arrived at by spectroscopic characterization and comparisons to known compounds, and chemical transformations. Some compounds included in this genre' are: O-methylsterigmatocystin (*31*)[102], demethylsterigmatocystin (*32*)[103], dihydro-O-methylsterigmatocystin (*33*)[104, 105], 5-methoxysterigmatocystin (*34*)[60, 106], aspertoxin (*35*)[107–109], austocystins (*36*)[110, 111], versicolorins (*37*)[112, 113], and aversin (*38*)[60].

B. Synthetic Approaches to Mycotoxins

1 Introduction

Although the aflatoxins and related compounds have elicited a considerable amount of biological and pharmacological interest, only a relatively meager level of attention, in the area of synthesis, has been proffered to date. This is somewhat surprising in that there is a great deal of importance attached to these molecules due to their high biological activity and widespread occurence. Several excellent reviews covering chemical and synthetic methods applicable to these highly oxygenated species have appeared in recent years[28, 31, 114–116]. This discussion will take into account all of the published synthetic work on the mycotoxins heretofore presented, as well as certain approaches to the toxins which contain synthetic interest.

Consideration of the generalized aflatoxin structure represented in (*1*) leads one to the inescapable conclusion that at the core of the system is the A-ring in the form of a phloroglucinol nucleus, in which each of the pendant phenolic oxygen atoms is uniquely differentiated. Furthermore, two of the three carbon sites on this nucleus are differentially substitued with carbon moieties. This constitutes a potentially difficult synthetic circumstance in that not only must a high degree of substitution be provided for, but also a certain amount of regiochemical control must be exercised in the aromatic system. Secondarily, the four carbons of the furo-[2,3-b]benzofuran BC piece must be appended, in either a complete manner, or sequentially in carbon units. In either case, suitable functionality should be present to: (a) allow for the facile formation of the bisfuranoid BC section, (b) provide accessibility to the C 15-C 16 vinyl ether functionality if necessary and, (c) give access to a hydroxyl group, or its functional equivalent, at C 14 if this is desirable for certain series of mycotoxins substituted at this position. Ideally, the functionality at C 14 should be of a nature which would allow its ready conversion to either a hydroxyl or a hydride, thereby making the intermediate of multifunctional utility. Thirdly, the formation of the cyclopentenone coumarin rings DE, or other coumarin system, from an activated phenolic precursor and a suitable β-ketoester (or a derivative thereof) must be undertaken. Conceptually, this is precedented and should incur no extraordinary difficulties, save of course the possible lability of any sufficiently sensitive functional groups in the reactants to the reaction conditions. The establishment of the C 13-C 14 cis relative stereochemistry should present no problems[116, 117], since the alternative formation of the trans isomer is believed to be highly unlikely, for stability reasons.

With the criteria set above for the mycotoxin syntheses certainly being desirable, if not essential, it behooves one to carefully scrutinize the available literature in order to ascertain the scope and applicability of the synthetic approaches reported. The presentation of synthetic efforts containing sufficient inherent flexibility to allow their possible application to the construction of the aflatoxins, and related mycotoxins, is now presented beginning with model system studies.

2 Syntheses of Model ADE Systems

Because of the structural nature of the ADE systems, the most straightforward and obvious method to accomplish a model synthesis is by the implementation of a von Pechmann coumarin synthesis[118–131] on a suitable functionalized phloroglucinol derivative. These methods have indeed seen rather general utilization in their application to this type of model system synthesis.

a) 5,7-Dimethoxycyclopenteno(c)coumarin (*4*)

This compound (*4*) was prepared both by Holker[132], and Buchi et al.[55], in connection with studies done to help elaborate the structures of the B and G series of toxins. Condensation of phloroglucinol dimethyl ether (*39*)[133] with diethyl β-oxoadipate (sulfuric acid) gave the coumarin (*6*), after a hydrolysis-cyclization (phosphoric acid) sequence. Compound (*4*) was then able to be obtained by catalytic hydrogenation of (*6*).

Convenient preparation of (*4*)[55] also was accomplished by the von Pechmann condensation of phloroglucinol dimethyl ether (*39*)[133] with ethyl cyclopentanone-2-carboxylate (*40*) in acidic media.

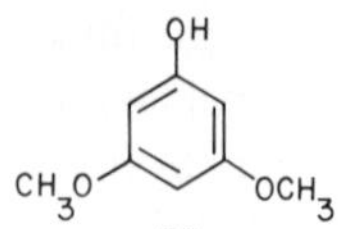

39

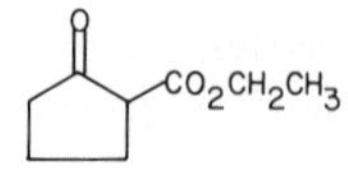

40

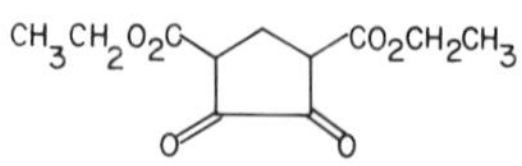

41

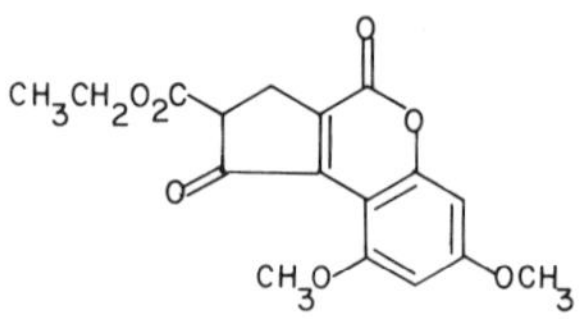

42

b) 5,7-Dimethoxycyclopentenon[3,2-c]coumarin (*5*)

Also used as a spectral comparison model for the elucidation of the B and G aflatoxins[55], this compound was prepared in a multi-step synthesis. The von Pechmann condensation of phloroglucinol dimethyl ether (*39*)[133] with diethyl cyclopentane-4,5-dione-1,3-dicarboxylate (*41*) in acidic solution afforded the β-ketoester (*42*), which readily underwent decarboalkoxylation, in a seperate step, to give the keto-coumarin (*5*). The beauty of this methodology is illustrated by the use of the symmetrical diketoester (*41*), which of course, only allows for the formation of a single coumarin (von Pechmann) product (*42*). The regiochemistry of the final product, however, was demonstrated to be the incorrect isomer insofar as the aflatoxin structures were concerned.

c) 5,7-Dimethoxycyclopentenon[2,3-c]coumarin (*6*)

The regiochemistry of this synthetic coumarin proved to be identical to that of natural aflatoxin B_1 (*1*)[132]. This preparation was also accomplished in two stages. The first proceeded by the acid catalyzed von Pechmann condensation of phloroglucinol dimethyl ether (*39*)[133] and β-ketoadipate ester (*43*)[134, 135] to give the substituted acyclic coumarin ester (*44*). This, in turn, underwent[55] cyclization to ketocoumarin (*6*) in polyphosphoric acid media.

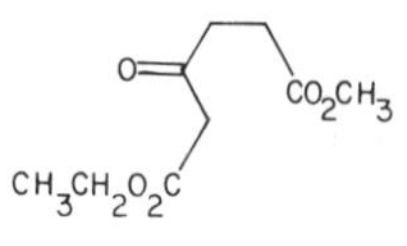

43

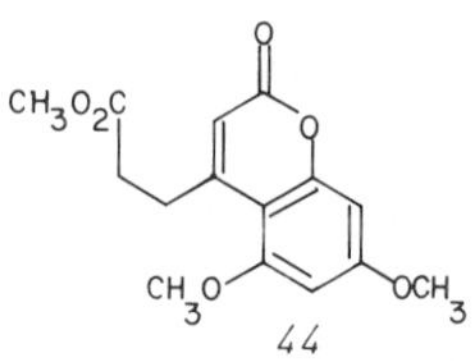

44

d) 5,7-Dibenzyloxyclopentenon[2,3-c]coumarin (*45*)

Although structurally similar to the model compounds previously discussed, a slight modification of the previous procedures was employed[136]. Von Pechmann conden-

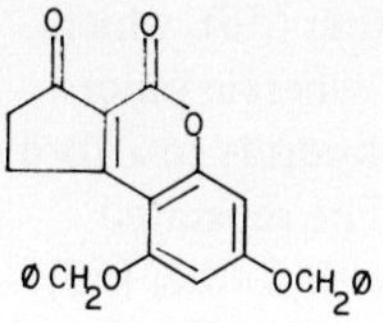

45

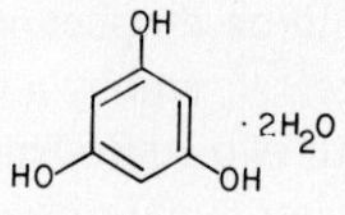

46

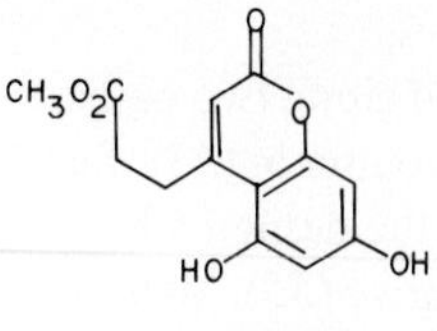

47

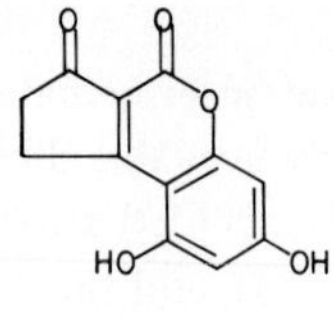

48

sation of phloroglucinol dihydrate (*46*) with ethyl methyl β-oxoadipate (*43*)[134, 135] under the conditions developed by Buchi[55], gave the substituted acyclic coumarin (*47*). Two methods of cyclization were employed, the result of both being the dihydroxy tricyclic material (*48*). Standard methods of benzyl ether formation were applied to the dihydric phenol to afford the desired dibenzyloxy coumarin (*45*).

The acute toxicity and carcinogenicity of the system were studied, the results demonstrating that (*45*) is neither toxic nor carcinogenic, in spite of its close structural relationship to aflatoxin B_1.

3 Aflatoxin Syntheses

a) Aflatoxins B_1 (*1*), B_2 (*8*), and B_{2a} (*12*)

The initial synthetic entry into the aflatoxin system was communicated by Buchi[66, 136], and reported the total synthesis of aflatoxin B_1 (*1*). The key elements of construction for this effort were, the selective differentiation of the phloroglucinol nucleus, and the formation of the functional precursor of the ABC system in the guise of the 4-methylcoumarin (*49*). Upon scrutinization of this molecule, it becomes evident that all of the requisite carbon atoms (see numbers) for the constitution of the ABC ring system are indeed present, albeit not necessarily in the proper level of oxidation. This fundamental coumarin piece (*49*), was arrived at by the implementation of two basic strategies.

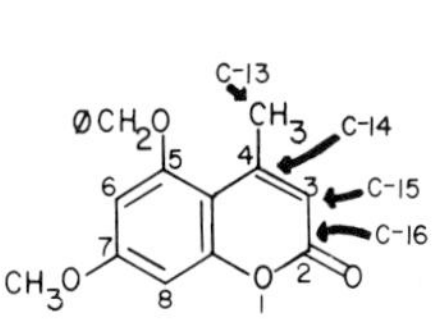

49

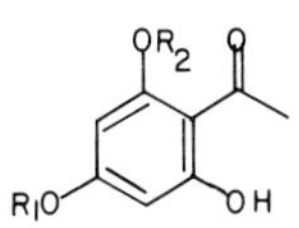

50 $R_1 = CH_3$ $R_2 = H$
51 $R_1 = R_2 = H$
52 $R_1 = CH_3$ $R_2 = CH_2\phi$

The first method began with phloroacetophenone-4-methyl ether (*50*), which was prepared by an established route[137], and by a newer method wherein phloroacetophenone (*51*) was acylated with two equivalents of acetic anhydride to afford a mixture of the 2,4 and the 2,6-diacetoxyphloroacetophenones. The separated 2,6-isomer was methylated and subsequently hydrolyzed to the methyl ether (*50*). Monobenzylation of (*50*) afforded benzyl ether (*52*), which underwent facile Wittig reaction with carbethoxymethylenetriphenylphosphorane to give the 4-methylcoumarin (*49*).

The second route to this key intermediate (*49*) also occured along two paths, both emanating from the readily available dihydroxy-4-methylcoumarin (*53*). Benzylation[115] of the dihydric phenol (*53*) afforded a mixture of the desired 5-benzyloxy-7-hydroxy substituted (*54*), and the dibenzylated derivative (*55*). Methylation of the monoprotected coumarin (*54*), then gave the desired (*49*).

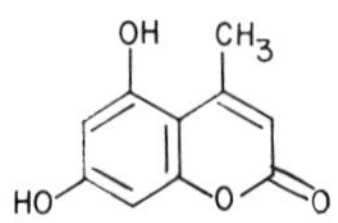

53

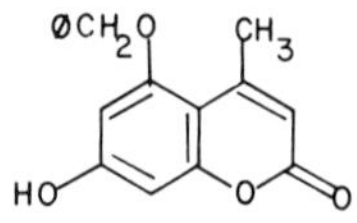

54

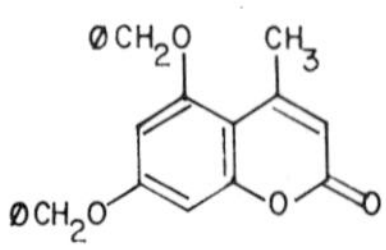

55

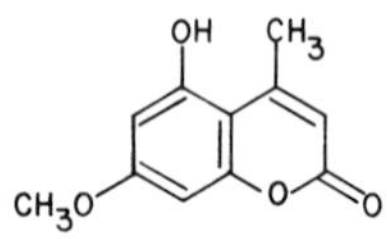

56

In the alternate approach, selective methylation[117, 138] of the dihydroxycoumarin (*53*) produced the 7-methyl ether (*56*) as the major product, which was readily benzylated to afford (*49*). In both of the above cases, the regiochemical nature of the compounds was ascertained by spectral comparisons with a known compound (*50*) and the compounds directly derived therefrom[137].

With the necessary carbons present in (*49*), it can be seen that the scheme of oxidation of the allylic methyl group to the aldehyde level (selenium dioxide), and a reduction of the coumarin with concommitant hydrolysis (zinc/acetic acid), afforded the hydroxy-aldehyde-acid intermediate (*57*), which expectedly and spontaneously cyclized to the tricyclic lactone (*58*) under the conditions of the reaction.

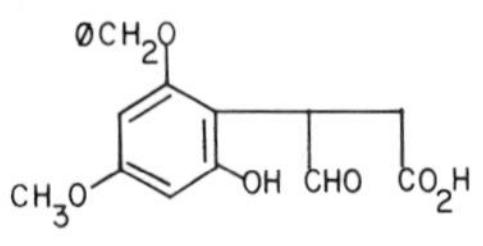

57

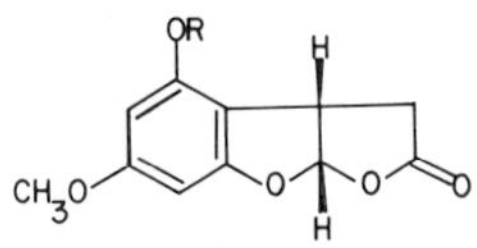

58 R = $CH_2Ø$

59 R = H

Thus, the prepared system (*58*) contains the necessary rudiments for the eventual construction of aflatoxin B_1 (*1*); namely, the regiochemically differentiated phloroglucinol system, and provisional functionality for the introduction of the vinyl ether moiety, both present in the fused ABC ring system.

It was felt that the lability of the enol ether was such that it could be adversely affected by the known conditions (vide infra) necessary to effect a von Pechmann condensation for the DE ring system fusion. Therefore, it was allowed that the introduction of this sensitive site would constitute the final step of the synthesis.

Catalytic debenzylation of (*58*) occured readily and the thus liberated phenol (*59*) treated with the β-ketoester (*43*) under von Pechmann conditions (methanolic HCl), which allowed for the formation of the annulated lactone hydrolysis product (*60*). The pentacyclic aflatoxin like system (*63*) was subsequently derived by acetal

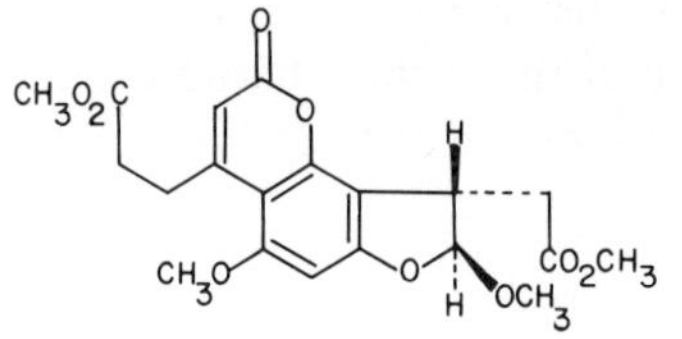

60

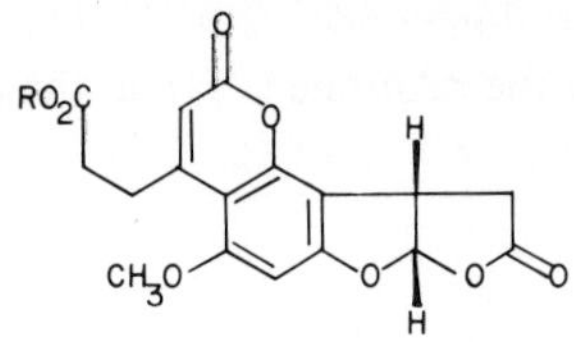

61 R=H
62 R=Cl

hydrolysis to lactone (*61*), followed by Lewis Acid catalyzed ($AlCl_3$) cyclization of the derived acid chloride (*62*). Conversion to the racemic aflatoxin B_1 (*1*) was realized by hydride reduction of the γ-lactone (disiamylborane) to hemiacetal (*64*) [aflatoxin B_{2a} = (*11*), which was shown to be identical by spectral means to that produced from the natural aflatoxin B_1 (*1*), by acid catalyzed hydration], and pyrolysis of the derived acyl acetal (*65*)[56].

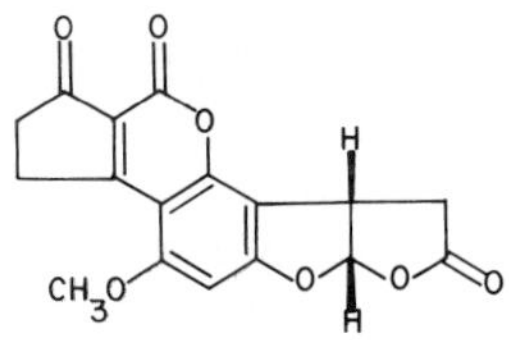

63

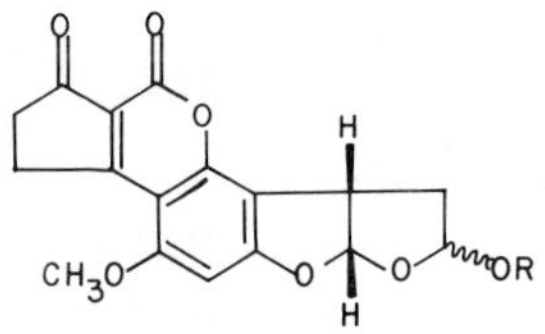

64 R=H
65 R=$COCH_3$

Thus, the total syntheses of the racemic aflatoxins B_1 (*1*), B_2 (*8*) [since this can be prepared by the controlled catalytic hydrogenation of aflatoxin B_1 (*1*)[57, 63]], and B_{2a} (*11*) were accomplished in a rather elegant carbon rearrangement fashion. The primary disadvantage of this route is manifest in the only moderate yields, and the regioisomeric mixtures prevalent in some of the early steps, pertaining to the differentiation of the phloroglucinol core.

A later refinement and modification of this route was communicated[139], wherein a new variation of the von Pechmann reaction was employed. The differentiated lactone (*58*) was converted to the acyl acetal (*66*) by a reduction (diisobutylaluminum

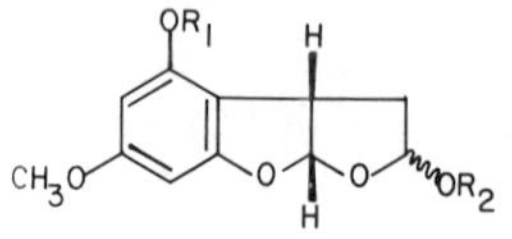

66 $R_1=CH_2Ø$ $R_2=COCH_3$
67 $R_1=H$ $R_2=COCH_3$
68 $R_1=R_2=COCH_3$

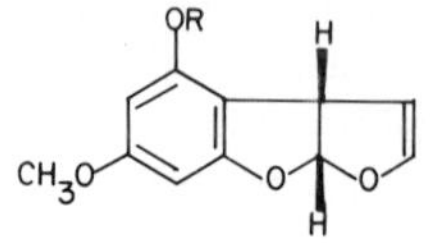

69 $R=COCH_3$
70 $R=H$

hydride), and acylation sequence. Catalytic debenzylation gave phenol (*67*), which upon acetylation afforded (*68*). The diacetate (*68*) underwent pyrolysis to vinyl ether (*69*), and further, hydrolysis of the phenolic acetate (potassium carbonate/water) produced the required phenol (*70*). Von Pechmann condensation of this very sensitive system was done under rather mild conditions, using the β-bromoenone (*71*)[139, 140] [from the diketone (*72*)], in the presence of zinc carbonate and sodium

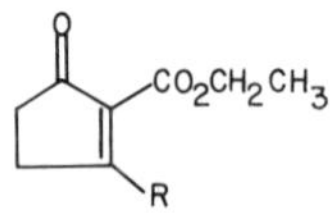

71 R = Br
72 R = OH

bicarbonate. Although the yield of the resulting racemic aflatoxin B_1 (*1*) can be considered only to be moderate (36%), the development and use of these conditions now allows more temporal latitude with regards to the introduction of the vinyl ether in the molecule. Thus, this method comprised a very valuable addition to syntheses of the type requiring a mild method of coumarin formation.

b) Aflatoxin B_2 (*8*)

The first reported construction of an intact ABC ring system residue generated for aflatoxin synthesis was described by Roberts[95, 117], and was an optically active compound (*73*), obtained as one of the degradation products of sterigmatocystin (*30*). A somewhat earlier report[117] of the racemic synthesis of (*73*) appeared which shortly preceeded the reported[66, 136] total synthesis of aflatoxin B_1 (*1*). Conceptually, the Buchi route is somewhat similar to this approach in that both utilize the 4-methyl coumarin analog of a differentiated phloroglucinol system to provide for the four carbons, and two of the oxygens, present in the furobenzofuran element of the aflatoxins.

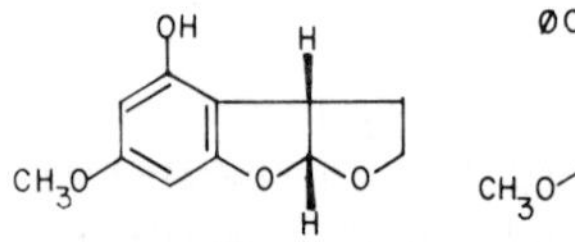

73

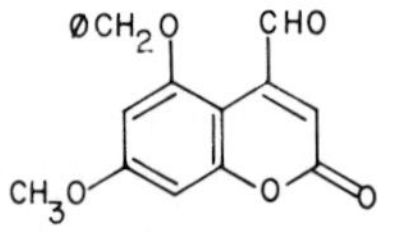

74

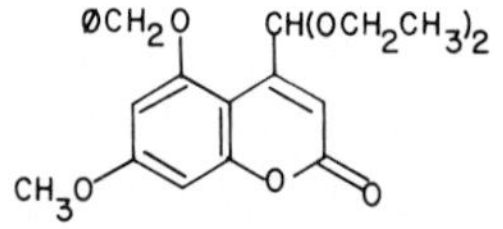

75

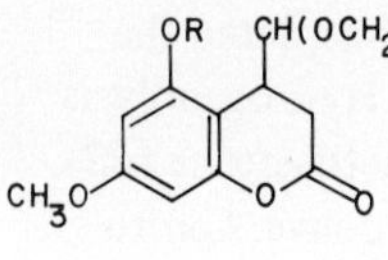

76 R=H

77 R=$COCH_3$

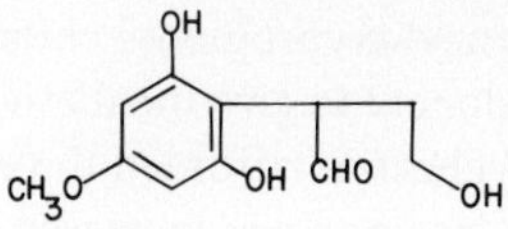

78

Dihydroxycoumarin (*53*) underwent selective methylation (dimethyl sulfate) to the 7-methoxy derivative (*56*), which upon benzylation and oxidation (selenium dioxide), afforded the 4-formyl coumarin (*74*). Conversion to the acetal (*75*) occured upon treatment with triethyl orthoformate, and subsequent catalytic hydrogenation served the dual purpose of removal of the benzyl group and reduction of the coumarin double bond, to give (*76*). Hydride reduction of the derived acetate (*77*), followed by acidic workup, gave directly the furobenzofuran (*73*) [presumably through the hydroxy aldehyde (*78*)]. Comparison of the spectra of racemic (*73*) with those of the naturally derived material showed the compounds to be identical.

The use of this tricyclic intermediate (*73*) to prepare a pentacyclic substrate in the form of tetrahydrodeoxoaflatoxin B_1 (*2*) was next communicated[141, 142)]. The condensation of phenol (*73*) with 2-carbethoxycyclopentanone (*40*) directly afforded (*2*), which was proven to be identical to the tetrahydrodeoxo compound prepared from natural aflatoxin B_1 (*1*).

Finally, this tricyclic phenol (*73*) was utilized[142)] to synthesize the natural metabolite aflatoxin B_2 (*8*) by the condensation with diethyl β-keto adipate, in von Pechmann fashion, to afford the acyclic ester (*79*). Upon treatment of the derived acid chloride (*80*) with Lewis Acid ($AlCl_3$), there was obtained a single product which

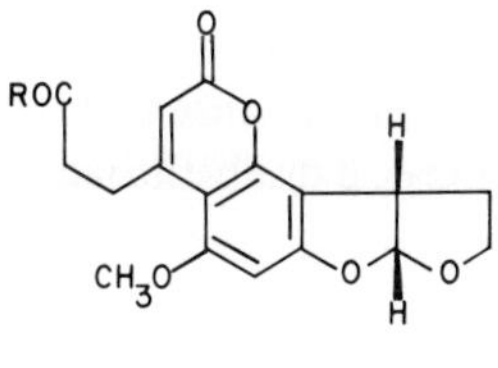

79 R=OCH_2CH_3

80 R=Cl

was spectroscopically and chromatographically identical with natural aflatoxin B_2 (8). Racemic tetrahydrodeoxoaflatoxin B_1 (*2*) was also prepared by catalytic hydrogenation of racemic aflatoxin B_2 (*8*). Although the difficulty of separation of the regioisomeric mixtures is avoided in this approach, the problem of a rather marginal yield in the von Pechmann reaction is still evidenced.

c) Aflatoxin G_1 (*9*)

As another example to test the applicability of the new variant of the coumarin synthesis that was developed, Buchi[139)] and Weinreb investigated a β-bromo carbonyl compound which would lead to aflatoxin G_1 (*9*), upon fusion with an appropriately substituted phenol. To this end, β-benzyloxypropionyl chloride[143] acylated the ethoxymagnesium salt of diethyl malonate to give the diester (*81*). Hydrogenolysis of the benzylic ether, followed by cyclization afforded the enolic ketolactone (*82*). As before, the electrophilicity of the position was increased by the conversion to the β-bromo carbonyl compound ($C_2O_2Br_2$) (*83*).

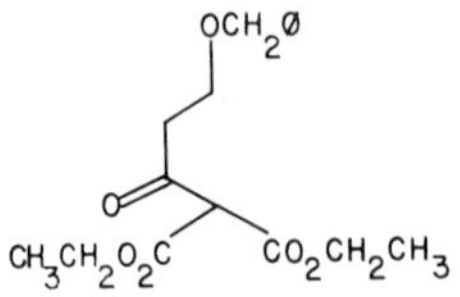

81

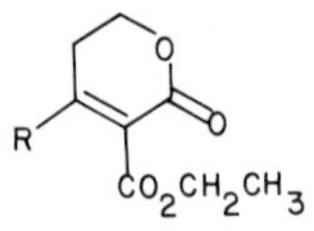

82 R = OH
83 R = Br

The phenolic constituent required for a synthesis of aflatoxin G_1 (*9*) is the same as that previously employed in the B_1 (*1*) synthesis, namely (*70*). Indeed, from the reaction of phenol (*70*) and vinyl bromide (*83*), in the presence of zinc carbonate and lithium iodide (a minor modification designed to increase the electrophilicity even more), there was obtained a modest amount of racemic aflatoxin G_1(*9*), which was spectroscopically shown to be identical with the natural material.

d) Aflatoxin G_2 (*10*) Studies

An interesting entry focusing on a synthetic approach to aflatoxin G_2 (*10*) was displayed by Roberts and co-workers[144]. The basis of this approach centered upon a method of construction of the DE coumarin lactone sector by the cyclization of functional appendages on the α and β positions (3 and 4) of the coumarin. Initial success was realized in the case of the model acetophenone (*84*), in the sense that Reformatsky reaction with diethyl bromomalonate afforded the α, β-disubstituted

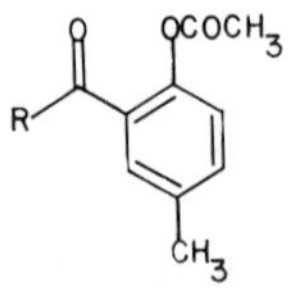

84 $R = CH_3$
85 $R = CH_2CH_3$

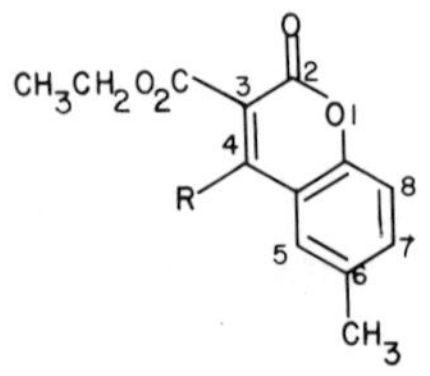

86 $R = CH_3$
87 $R = CH_2CH_3$

coumarin (*85*). However, extension of this methodology to the necessary propiophenone (*86*) caused a dramatic reduction in the yield of the corresponding coumarin (*87*), presumably due the increased steric interaction effects.

In an effort to circumvent this difficulty, a situation encompassing an intramolecular aldol condensation was envisioned. To this end, the model methoxypropiophenone (*88*) was acylated with malonyl monoacid chloride mono ethyl ester, and

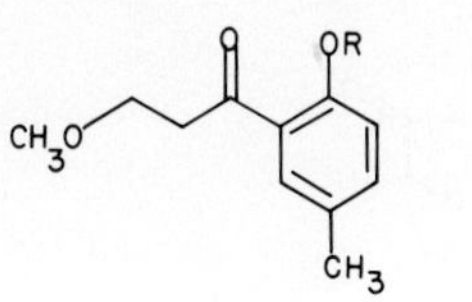

88 R = H

89 R = $COCH_2CO_2CH_2CH_3$

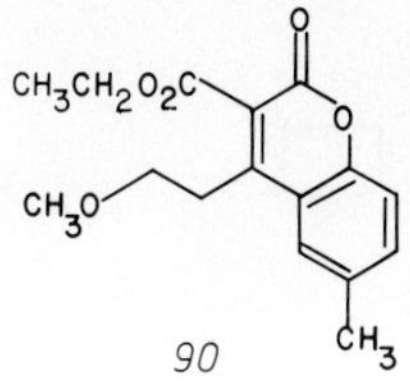

90

the resulting ketodiester (*89*) cyclized intramolecularly, to give the necessary α,β-substituted coumarin (*90*). Treatment with acid (sulfuric acid) caused lactonization to the coumarin lactone (*91*).

Unfortunately, modification of the aromatic nucleus to a phloroglucinol dimethyl ether system (*92*) adversely affected the ability of the phenolic hydroxyl to undergo acylation. This was found to be inconsequential however, since the acylated product derived from resorcinol monomethyl ether (*93*) was completely resistant to the intra-

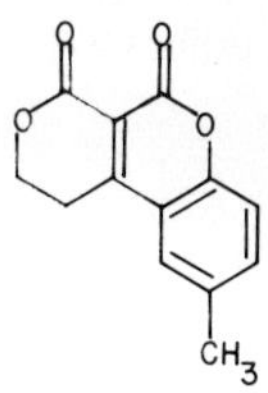

91

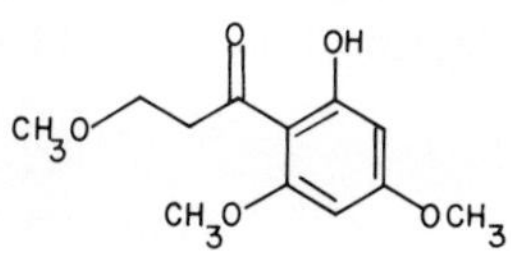

92

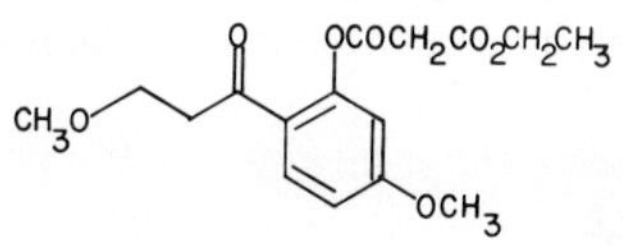

93

molecular cyclization step. This was attributed to the high electron releasing capability of the methoxy group situated para to the propiophenone carbonyl, which makes this site much less susceptible to nucleophilic attack by the β-ketoester appendage.

e) Aflatoxin M_1 (*13*) and M_2 (*14*)

The desirability of having relatively copious quantities of the "milk toxins" for studies regarding their carcinogenicity, prompted Buchi and Weinreb[139, 145] to develop a total synthesis of these metabolites. Notably, the only difference between aflatoxin M_1 (*13*) and aflatoxin B_1 (*1*) is the presence of the tertiary and benzylic hydroxyl functionality at C-14. Unfortunately, the methodologies utilized in the constructions of the toxins previously discussed are not applicable to this problem in that they inherently do not contain the necessary provisions for the initial presence, or the delayed introduction, of this C-14 hydroxyl. Therefore, a radically different approach was essential, which would provide for this contingency.

The basic strategy for the formation of the final pentacyclic system via the von Pechmann condensation of an intact ABC system, and a suitable β-ketoester (or a suitable derivative thereof) was retained, since no obvious difficulty was evident based on prior experiences. Work now was concentrated on the synthesis of the quintessential tricyclic ABC system (*94*).

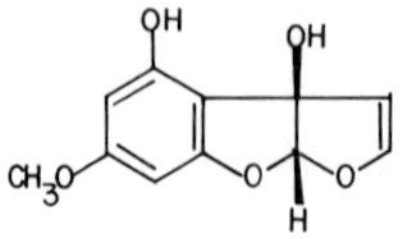

94

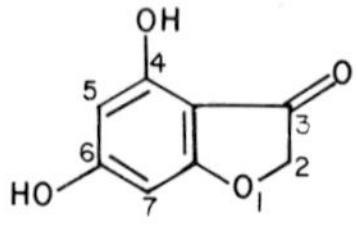

95

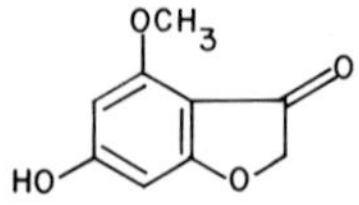

96

The pathway led through the successive annulations of the B, and then the C, rings onto the phloroglucinol nucleus, with the provision for phenol differentiation being introduced at a point along this sequence. The two carbons for the formation of the B ring were provided in the form of chloracetonitrile addition to phloroglucinol[146], producing the dihydric coumarin (*95*), already containing a C-3 carbonyl group as the functional precursor to the hydroxyl.

At this juncture, a series of investigations[139] concerning the relative reactivities of the phenolic functions were undertaken in order to form the basis for the regioselective generation of the 4-hydroxy-6-methoxy system. The information obtained provided the following facts: (a) selective methylation (diazomethane) of (*95*), gave the 4-methoxy ether (*96*) and, (b) selective deprotection of the dibenzyl ether (*97*)

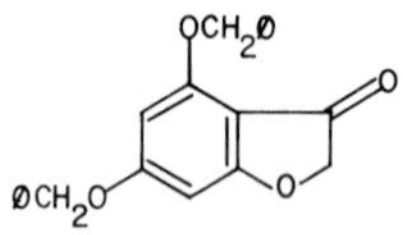

97

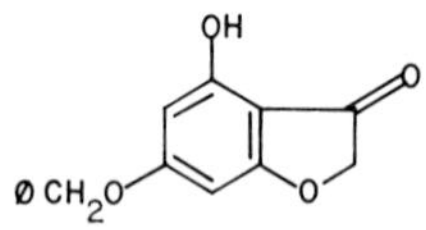

98

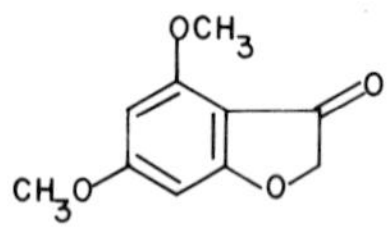

99

gave preferentially the 6-benzyloxy system (*98*). However, from the Lewis Acid ($AlCl_3$) cleavage of the dimethyl ether (*99*)[147] was obtained the desired C-4 phenol (*100*) (identified by comparison with an unambiguous sample[148]), which was benzylated to afford the differentially substituted coumarin (*101*). A protected oxygen

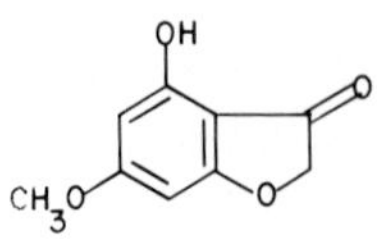

100

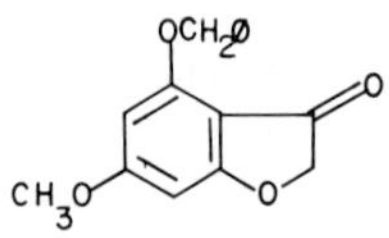

101

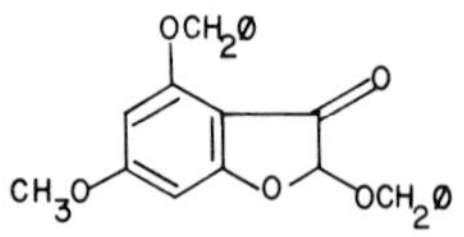

102

at C-2 (*102*) for use in the construction of the C-ring was introduced next by the sequence of bromination, and subsequent displacement with the anion of benzyl alcohol. Concurrent introduction of the C-3 alcohol, and a three carbon residue to be used as an acetaldehyde equivalent for C-ring annulation, was accomplished by the condensation of ketone (*102*) with allylmagnesium bromide, thereby giving access to alcohol (*103*), as an epimeric mixture. Oxidative cleavage (osmium tetroxide/sodium periodate) gave the corresponding epimeric aldehydes (*104*) and (*105*), that

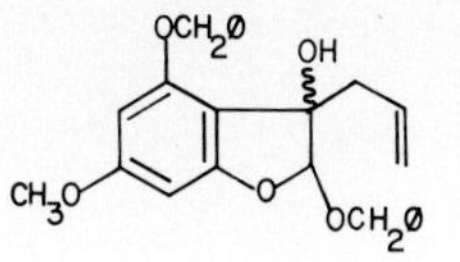

103

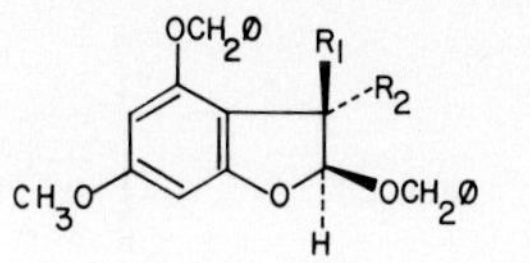

104 $R_1 = CH_2CHO$ $R_2 = OH$
105 $R_1 = OH$ $R_2 = CH_2CHO$

furnished the epimeric acetates, (*106*) and (*107*) upon selective catalytic debenzylation in an acylating medium. The diastereomeric mixture as then subjected to the further action of catalyst and hydrogen to yield exclusively the more

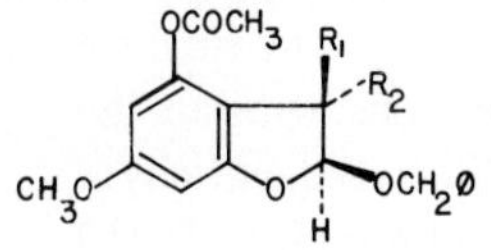

106 $R_1 = CH_2CHO$ $R_2 = OH$
107 $R_1 = OH$ $R_2 = CH_2CHO$

stable, cis fused, tricyclic hemiacetal (*108*). The acylacetal (*109*) was readily formed at low temperature (acetic anhydride/pyridine), which short term contact Kraft pyrolysis (450 °C) easily converted to vinyl ether (*110*). Mild hydrolysis of the phenolic

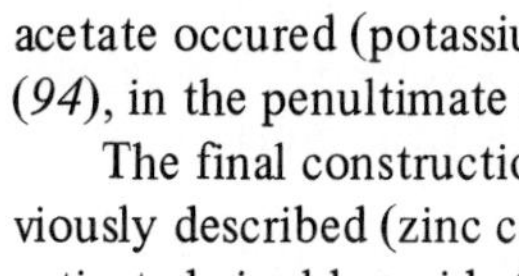

108 R = H
109 $R = COCH_3$

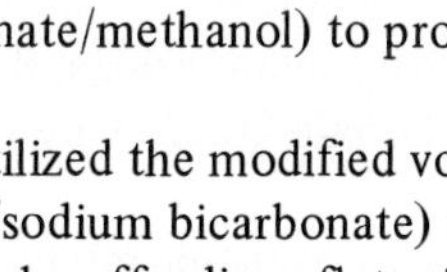

110

acetate occured (potassium carbonate/methanol) to produce the requisite phenol (*94*), in the penultimate step.

The final construction step utilized the modified von Pechmann conditions previously described (zinc carbonate/sodium bicarbonate) to fuse phenol (*94*) to the activated vinyl bromide (*72*), thereby affording aflatoxin M_1 (*13*). The route discussed above also presents a formal total synthesis of aflatoxin M_2 (*14*), since this is able to be prepared[76)] through the controlled catalytic hydrogenation of aflatoxin M_1 (*13*). It should also be possible to apply this technology in a synthesis of aflatoxin GM_1 (*15*), by the use of a suitable substrate [possibly (*83*)] in the von Pechmann reaction with phenol (*94*).

f) Aflatoxin P_1 (*18*)

The synthesis of aflatoxin P_1 (*18*) has been accomplished, although the details have not been reported in the literature[149)]. The acetylation of the dihydric coumarin

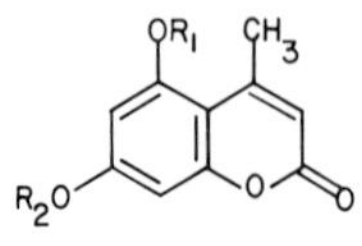

111 $R_1 = R_2 = COCH_3$

112 $R_1 = COCH_3$ $R_2 = H$

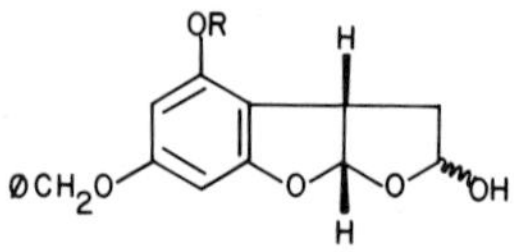

113 $R = COCH_3$

114 $R = H$

(*53*) gave the diacetate (*111*), which was selectively hydrolyzed (p-toluenesulfonic acid/methanol) to the 5-acetoxy-7-hydroxy coumarin (*112*). Benzylation of the phenol, oxidation (selenium dioxide) of the 4-methyl group to the aldehyde, and application of the familiar rearrangement conditions[66, 136] used previously, afforded the hemiacetal (*113*), which easily hydrolyzed to the phenol (*114*). The final step utilized the application of a rather large number of transformations, without the characterization of any of the intermediates. These steps included sequentially: (a) the von Pechmann condensation of phenol (*114*) and vinyl bromide (*72*), (b) acylation of the acetal at C-16, (c) hydrogenolysis of the benzyl ether, (d) acetylation of the free phenol, (e) vacuum Kraft pyrolysis to afford the vinyl ether and, (f) hydrolysis of the phenolic acetate, all utimately leading to the isolation of racemic aflatoxin P_1 (*18*), which was shown to be identical to natural aflatoxin P_1 (*18*) upon comparison of tlc, ultraviolet, and mass spectral data.

A partial synthesis of this metabolite has also been reported by Buchi and Wogan[84], proceeding to the optically active natural product (*18*) by means of demethylation (lithium alkyl mercaptides) of natural aflatoxin B_1 (*1*) in good yield.

Investigations of the acute toxicity of aflatoxin P_1 (*18*) vs aflatoxin B_1 (*1*) indicated that the former had less than 5% of the potency of the latter, in the assay systems studied.

g) Aflatoxin Q_1 (*25*)

In an effort to prepare sizable quantities of this liver metabolite for studies pertaining to structure vs biological activity relationships, Buchi and co-workers[150] developed two methods of preparing aflatoxin Q_1 (*19*), by the chemical transformation of natural aflatoxin B_1 (*1*).

The treatment of the readily available model cyclopenteno coumarin (*6*)[55] with silver (I) oxide/sodium hydroxide, of thallium (I) ethoxide/hydrogen peroxide, gave fair to good yields of the desired Q_1 analog (*115*). [Exhaustive hydrogenolysis of (*115*), prepared by either method, proceeded by absorption of three equivalents of

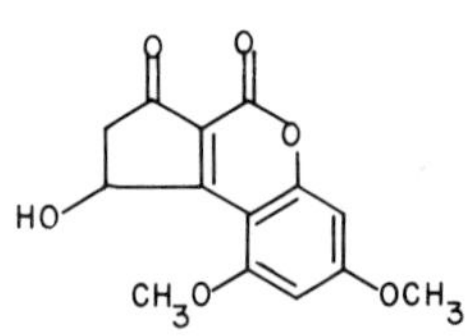

115

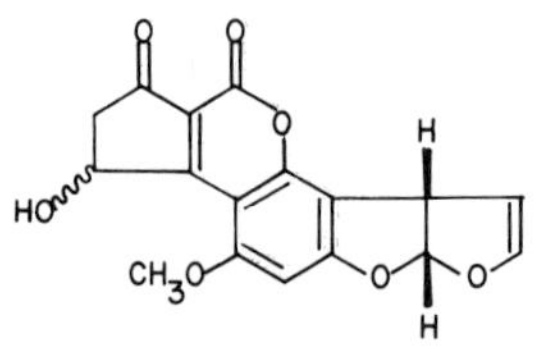

116 a mixture of diastereomers

hydrogen to produce (*4*)]. As expected, upon application of these same conditions to the desired substrate, natural aflatoxin B_1 (*1*) suffered smooth conversion to a mixture of diastereomers (*116*), which were separated and compared to natural aflatoxin Q_1 (*25*) (circular dichroism) in order to determine which diastereomer was identical to the naturally occuring one.

4 Syntheses of Related Metabolites

a) Dihydro-O-methylsterigmatocystin (*33*)

The first total synthesis of a sterigmatocystin type structure was reported by Roberts[151]. The furobenzofuran section[73] used in the construction was the same piece previously encountered[95, 117, 142] in the synthesis of aflatoxin B_2 (*8*). Ullman type coupling[152] (pyridine/cuprous chloride) with bromoester (*117*), and subsequent hydrolysis, gave the diphenyl ether (*118*). Intramolecular cyclization to form

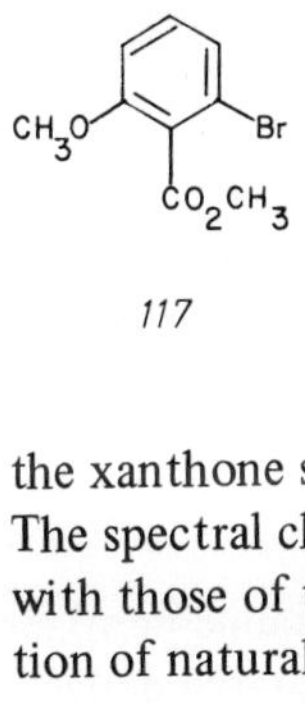

117

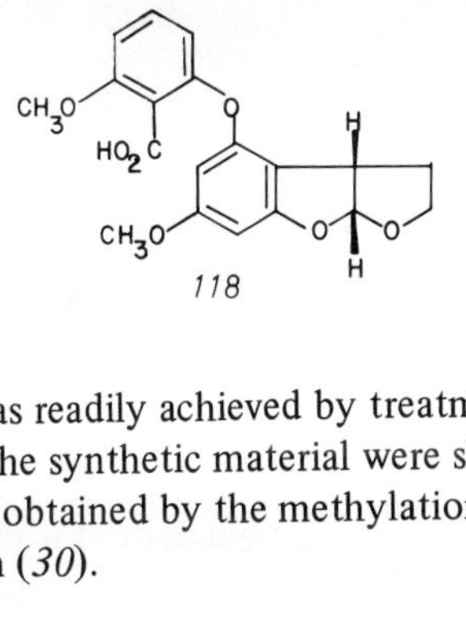

118

the xanthone system of (*33*) was readily achieved by treatment with oxalyl chloride. The spectral characteristics of the synthetic material were shown to be identical with those of the material (*33*) obtained by the methylation, and catalytic reduction of natural sterigmatocystin (*30*).

b) O-Methylaversin (*119*)

In order to help to illuminate the structures of a series of mold metabolites, O-methylaversin (*119*) was prepared[153]. Again, the furobenzofuran (*73*)[95, 117, 142] was the starting material. The synthesis commenced with the preparation of lactone (*120*), by the condensation of (*73*) with oxalyl chloride. Methanolysis of the lactone gave

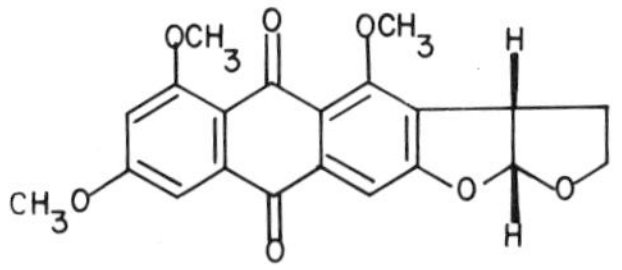

119

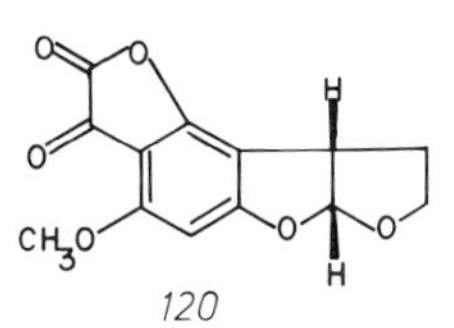

120

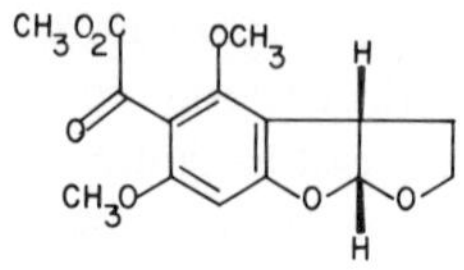

121

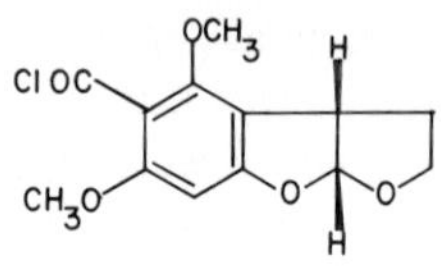

122

α-ketoester (*121*), that was converted to the desired acyl chloride (*122*) (*via* the α-ketoacid). The other portion of the aversin skeleton was added to (*122*) in the form of 3,5-dimethoxybenzonitrile[154, 155], by a Friedel-Crafts acylation, resulting in the ketone (*123*). Treatment of (*123*) with base induced cyclization[156, 157], to an unfortunate mixture (1:1) of anthrone precursors (*124*) and (*125*). Separation

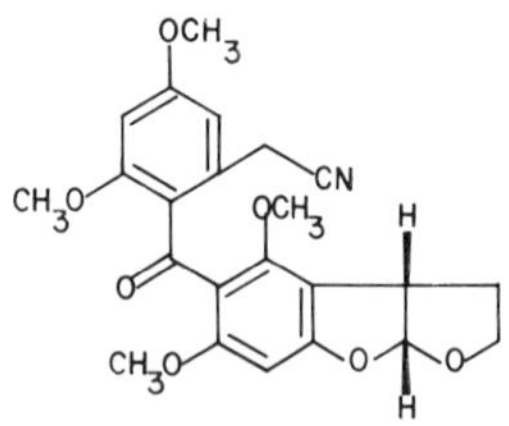

123

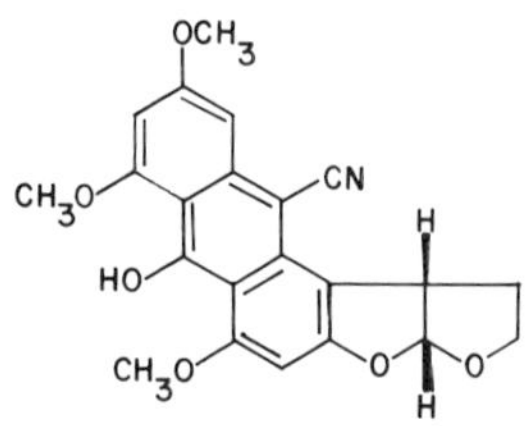

124

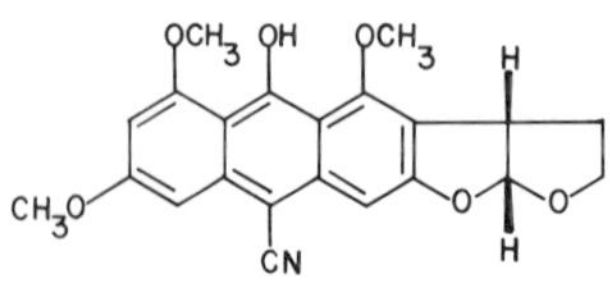

125

of the regioisomeric mixture, and treatment of isomer (*125*) with alkaline peroxide, afforded O-methylaversin (*119*), which was spectrally identical to material derived from natural sources. This synthetic sequence provided absolute proof of the structure of aversin (*38*).

The formation of the regioisomeric products (*124*) and (*125*) in the cyclization step, proves to be the only serious problem in this otherwise excellent synthetic effort.

c) O-Methylsterigmatocystin (*31*)

Roberts synthesis of O-methylsterigmatocystin (*31*)[158] embodied the same xanthone type synthesis as previously encounterd in the dihydro derivative (*33*). Provision for the introduction of the vinyl ether, however, was accomplished through the use of a lactone carbonyl at C-2.

The known phenolic lactone (*59*)[66, 136] was ring opened (methanol/hydrogen chloride) to acetal (*126*), which subsequently underwent Ullman type reaction[151, 152] and acid catalyzed lactonization, to yield lactone (*127*). Conversion to the corre-

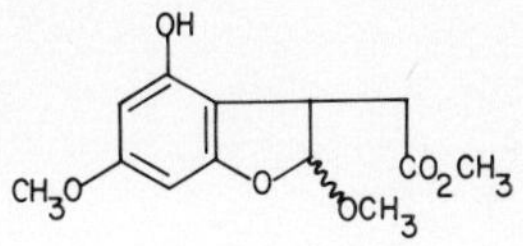

126

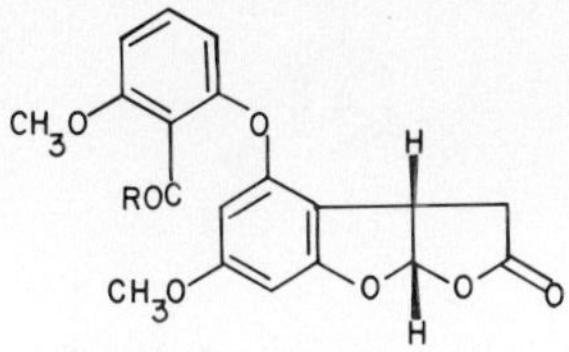

127 R=OCH$_3$
128 R=Cl

sponding acid chloride (*128*), and cyclization, achieved formation of the pentacyclic sterigmatocystin skeleton (*129*). Introduction of the vinyl ether was accomplished as previously described by Buchi[66, 136], through conversion of the γ-butyrolactone to the acylacetal (*130*) (hydride reduction; acetylation), and repeated exposure to

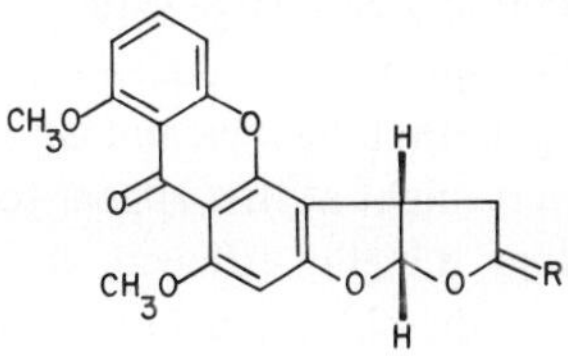

129 R=O
130 R=H, COCH$_3$

the technique of Kraft pyrolysis, thereby giving racemic O-methylsterigmatocystin (*31*). This was proven to be spectroscopically and chromatographically identical to an authentic sample obtained from natural sources.

5 A New Approach to Aflatoxin Synthesis-Target, Aflatoxins M_1 (*13*) and M_2 (*14*)

A very recent approach (work is currently in progress) by Buchi and co-workers to the aflatoxin skeleton deserves mention at this juncture. It was concieved as an effort by which sizable quantities of the M series of toxins (tertiary and benzylic hydroxyl) would become attainable for carcinogenicity studies, by a terse, high-yield synthetic route.

The previously announced route[139, 145] to aflatoxin M_1 (*13*), made use of an:
1) A ⟶ AB ⟶ ABC;
2) ABC + E ⟶ ABCDE

approach. In the proposed new route, the second stage condensation of the intact ABC and E systems in von Pechmann fashion to form the pentacyclic aflatoxin system, is envisioned to be acceptable as previously presented[139, 145]. The concentration of this synthesis focuses upon a novel and efficient approach to a differentiated furobenzofuran system (*131*), via the condensation of a suitably differentiated phloroglucinol nucleus A, (*132*), (X and Y are different

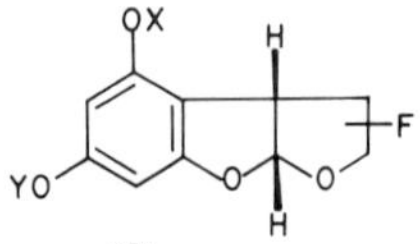

131

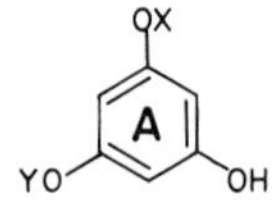

132

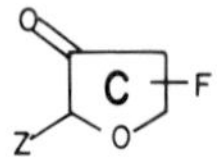

133

protecting groups), with a functionalized (F = functionality at either the C-15 or C-16 position which would allow latent access to the enol ether between these two carbon atoms) 3-oxofuran moiety C, (*133*)[159–161]. This, in essence, would comprise a new type of synthetic logic for the first stage, in the sense of an A + C ⟶ ABC spproach.

The A-ring is prepared[162] in very facile fashion, by applying the observations of Kampouris[163, 164], that allowed for the formation and the selective basic hydrolysis of the polyarylsulfonate esters of polyhydric phenols. In combination with both standard, and more recent innovative[165] techniques of ether formation, and taken in the proper sequence, these results lead to a potentially very large variety of phloroglucinol substitution patterns. In practice, a wide spectrum of differentially protected phloroglucinol entities were prepared, utilizing methyl, benzyl, and allyl ether combinations for both X and Y in (*132*). Although it might at first appear to be a rather laborious road to the A piece, this methodology is highly efficient, and lends itself readily to the preparation of large quantities of the phloroglucinol derivative.

The C portion was originally projected to be a protected acetal of a 3-oxofuran such as (*134*). However, a much more accessible compound was attainable in the

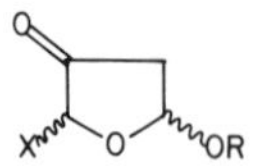

134

form of (*135*) (X = Br). The preparation of this compound commences from 1,4-anhydroerythritol (*136*)[166], by selective differentiation of the cis-hydroxyls as the monoacetate (*137*) (triethyl orthoacetate/acid; aqueous oxalic acid). Oxidation then affords the 3-oxofuran (*138*) (X = H), which undergoes free radical bromination (N-bromosuccinimide) to give the bromofuranone (*135*) in good overall yield.

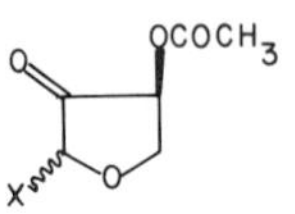

135 X = Br
136 X = H

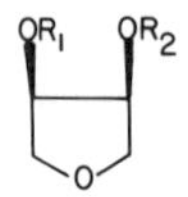

137 $R_1 = R_2 = H$
138 $R_1 = H$ $R_2 = COCH_3$

Fusion in the example shown for phloroglucinol methyl benzyl ether (*139*) and bromofuranone (*135*) does indeed occur to afford, not unexpectedly, approximately a 1:1 mixture of the regioisomeric furobenzofuran bromides (*140*) and (*141*) (presumably formed by the reaction of the tertiary-benzylic alcohol of the initial condensa-

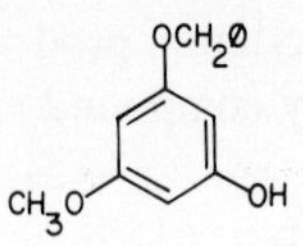

139

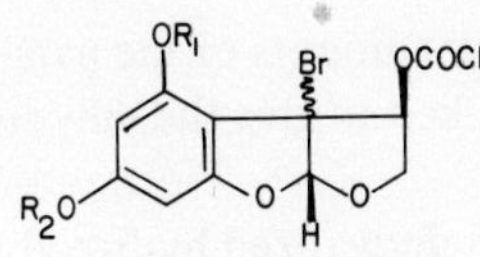

140 R_1=CH_2Ø R_2=CH_3

141 R_1=CH_3 R_2=CH_2Ø

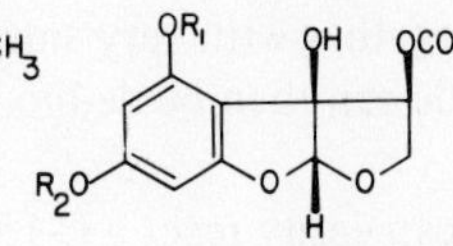

142 R_1=CH_2Ø R_2=CH_3

143 R_1=CH_3 R_2=CH_2Ø

tion product with the liberated nascent hydrogen bromide). The mixture is solvolyzed to the mixture of alcohols (*142*) and (*143*) and separated at this point.

The relative stereochemistry of the three asymmetric centers was determined to be cis by virtue of both; (a) the ready and extremely rapid acetylation of both alcohols in the derived diols (*144*) and (*145*) and, (b) facile and quantitative formation of the acetonide between the vicinal hydroxyls to give (*146*) and (*147*). Further-

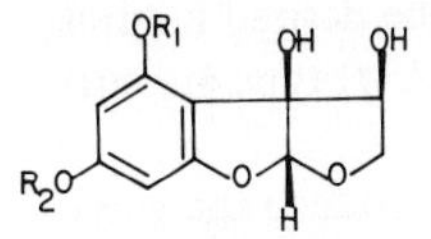

144 R_1=CH_2Ø R_2=CH_3

145 R_1=CH_3 R_2=CH_2Ø

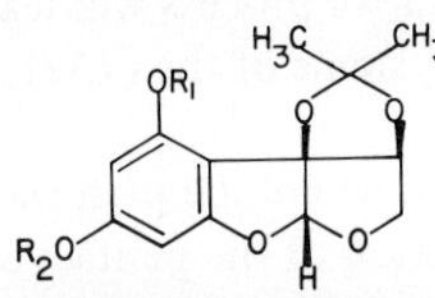

146 R_1=CH_2Ø R_2=CH_3

147 R_1=CH_3 R_2=CH_2Ø

148 R_1=H R_2=CH_3

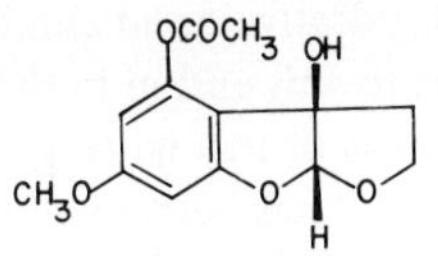

149

more, the correct regioisomer was identified by spectral comparison of an authentic sample of (*148*)[139, 145] with the lateral compounds derived by chemical manipulations of both (*142*) and (*143*).

Thus, in addition to the intact basic ring structure of the ABC system, one has attained both regioisomeric differentiation of the phenol necessary for the von Pechmann condensation, and provision for the vinyl ether insertion in the form of the C-15 hydroxyl (or a possible derivative thereof). Indeed, it is evident that any of several directions may be taken in this route by variations in: (a) the nature of the functional groups on the phloroglucinol nucleus and C-15 hydroxyl, (b) the timing of their introduction and removal and, (c) the timing for the introduction of the vinyl ether moiety.

One such alternative pathway is presented in the scheme wherein the introduction of the vinyl ether is postulated as the final step. Using modified von Pechmann conditions, the condensation of phenol (*149*) (prepared from (*146*) by catalytic hydrogenolysis) with vinyl bromide (*71*) afforded the pentacyclic acetonide (*150*)

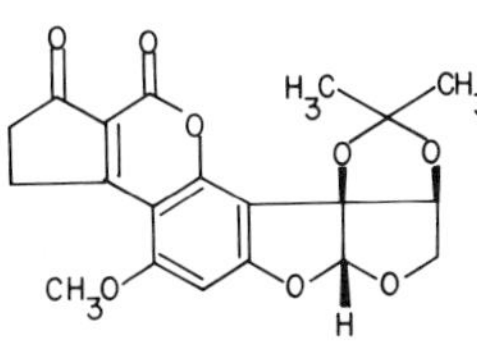

150

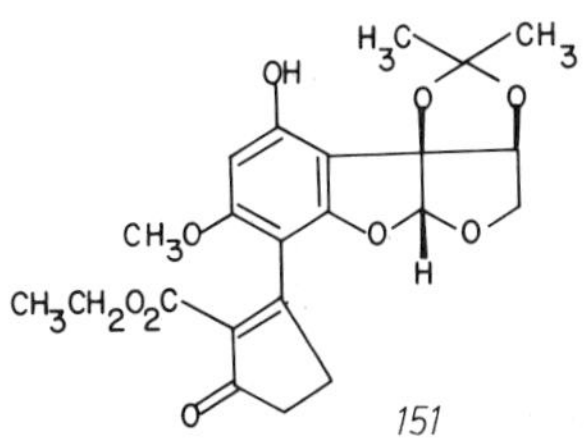

151

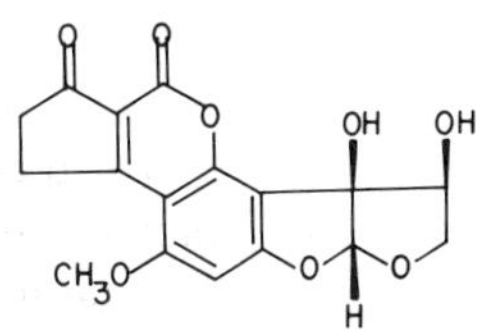

152

in good overall yield, together with very small amounts of the para-alkylation product (*151*). The acetonide can then be de-blocked to give the dihydroxy compound (*152*).

Alternatively, one can easily react a C-15 derivatized hydroxyl of type (*153*) directly, to afford an intermediate of type (154). In either circumstance, it is ob-

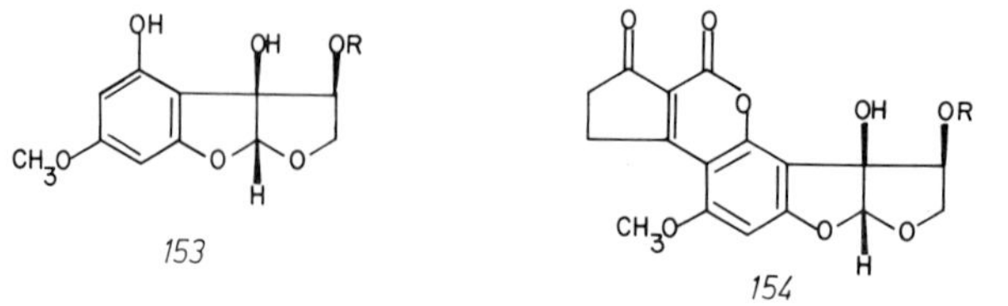

153 154

served that a dehydrative or an eliminative process will lead to the desired natural product. Studies to this end in both systems of the (*152*), and (*154*) type are currently in progress as of this writing.

Yet a third road has been explored where the plan was to introduce the vinyl ether functionality prior to the formation of the pentacycle. This is accomplished via the use of the allyl protecting group in the phloroglucinol nucleus (*155*), leading eventually to the tricyclic acetate (*156*). This is easily transformable to the tosylate (*157*), which suffers elimination (diazabicycloundecene) in rather dissapointing yield, to give the vinyl ether (*158*)[167]. Selective removal of the phenolic allyl ether

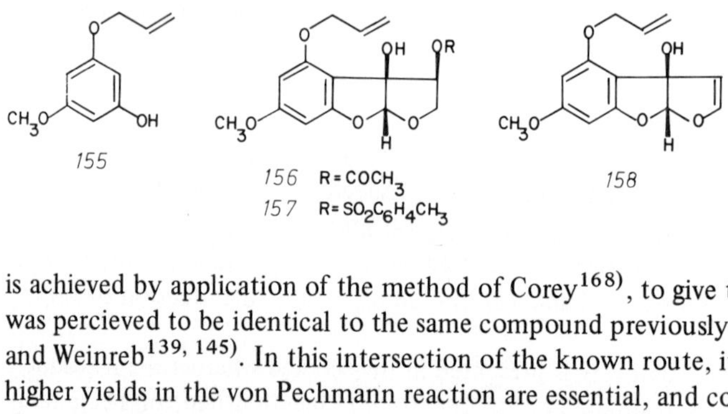

155 156 R = $COCH_3$ 157 R = $SO_2C_6H_4CH_3$ 158

is achieved by application of the method of Corey[168], to give the phenol (*94*), which was percieved to be identical to the same compound previously prepared by Buchi and Weinreb[139, 145]. In this intersection of the known route, it is observed that higher yields in the von Pechmann reaction are essential, and conditions to improve the status of this problem are currently under scrutiny.

The negative aspects of this route are seen as twofold. Firstly, the yield of the condensation product between the phloroglucinol derivative and the bromofuranone (*135*) tends to be somewhat low, even though they appear to give only one isolable coupling product. Secondly, the fact that regioisomeric mixtures are necessarily formed in this step when an unsymmetrical phloroglucinol piece is used, is certainly undesirable, even though the incorrect regioisomer may be useful in the synthesis of other families of toxins.

However, positively speaking, this method lends itself readily adaptable to the reasonably large scale preparation of intermediates, and allows for a wide amount of temporal and functional latitude with regards to the synthetic route. Also, many of the steps in the body of the synthesis are relatively trivial, and high yield (e.g. hydrolysis, tosylation, debenzylation). In addition, it should be noted that this methodology might also hold promise for the synthesis of the C-14 hydrido series

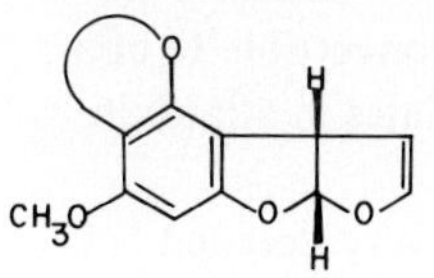

159

of toxins [see (*159*)] by use of an appropriate reduction at the stage of the benzylic bromides [e.g. (*139*) and (*140*)], thus adding a possible extra measure of versatility to this new synthetic path.

6 Biosynthesis of Mycotoxins

Much speculation regarding the biosynthesis of these metabolites has appeared[169]. The widespread usage of ^{13}C nuclear magnetic resonance techniques has greatly assisted in the confirmation, clarification, or rejection of many of the postulated routes.

Very recent studies have indicated that an essential ingredient in aflatoxin biosynthesis was the fungal pigment averufin (*160*)[170, 171]. Indeed, it was also shown[172–174] that the bisfuran ring system in aflatoxin was concieved by the rearrangement[175–177] of the C-6 averufin side chain, rather than by fission, followed

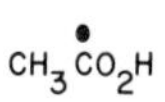

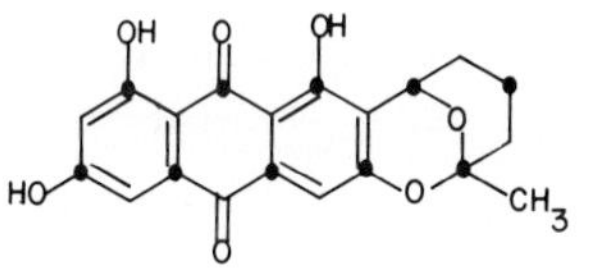

160

161

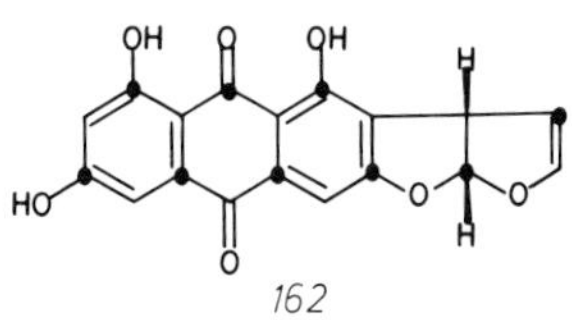

162

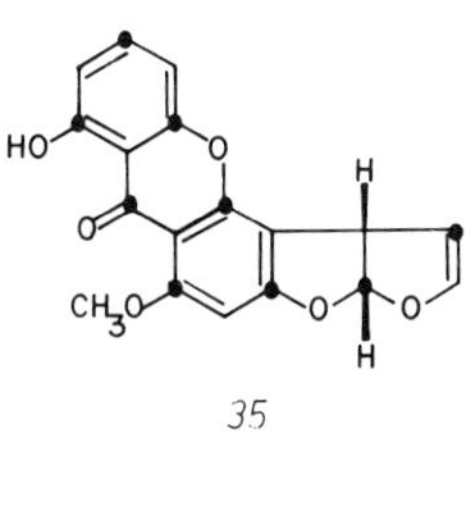

35

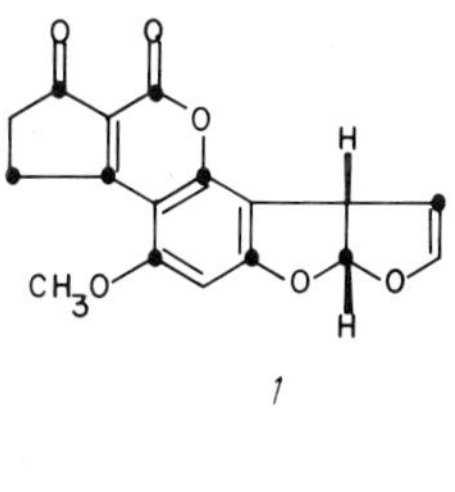

1

by the reinstatement of the bisfuran through an acetoacetate moiety. This was implied by the fact that when ^{13}C labelled averufin (*160*) (from labelled acetate) is converted to aflatoxin B_1 (*1*), the labelling pattern was the same as that resulting from the direct formation of aflatoxin B_1 (*1*) from ^{13}C acetate. Additional work has determined that besides averufin (*160*), norsolorinic acid[178], versiconal acetate[179], versicolorin A[180], and sterigmatocystin[181, 182], are convertible to aflatoxin B_1 (*1*), and therefore might occupy positions as intermediates in aflatoxin biosynthesis.

The relationship of ^{13}C acetate and averufin (*160*)[183] and very detailed ^{13}C nuclear magnetic resonance enrichment studies of the abovementioned intermediates have recently led Steyn and co-workers[184–187] to conclude[187] that a biosynthetic pathway starting with polyketide (*161*), going through averufin (*160*), versicolorin A(*162*), and sterigmatocystin (*30*), eventually results in aflatoxin B_1 (*1*).

C. Conclusions

As the general populace becomes more concerned with environmental and health hazards, detailed investigations into the physical, physiological, analytical, and chemical nature of these mycotoxins becomes increasingly imperative. Sizable contributions in the areas of isolation, biological activity, and structural elucidation have been mentioned. Additionally, the truly excellent synthetic efforts of both Buchi, and Roberts to construct the carbon skeletons of mycotoxins, has led to several total syntheses of mold metabolites. Further chemical and synthetic efforts are currently under investigation in order to more rigorously define the reactivity of the mycotoxin systems to different reagents, in the hope that these studies might more clearly delineate the role of these compounds in toxicosis and/or carcinogenesis[11, 12, 26, 188].

Acknowledgements. The author would like to express his sincere gratitude and appreciation to Professor George Büchi of The Massachusetts Institute of Technology for his encouragement and direction in studies done on the aflatoxins, and for his kindness in extending support in the form of a post-doctoral fellowship in his laboratories. His display of great interest and knowledge serves as a source of inspiration to all of his students and co-workers.

The author would also like to express thanks to PPG Industries Inc., Pittsburgh, Pennsylvania for the use of library facilities during the preparation of part of this manuscript.

D. References

1. Allcroft, R., Carnaghan, R. B. A., Sargeant, K., O'Kelly, J.: J. Vet. Rec. *73*, 428 (1961)
2. Delongh, H., Beerthuis, R. K., Vles, R. O., Barnett, C. B., Ord, W. O.: Biophys. Acta. *63*, 548 (1962)
3. Kraybill, H. F., Shimkin, M. B.: Adv. Cancer Res. *8*, 191 (1964)

4. Sargeant, K., Carnaghan, R. B. A., Allcroft, R.: Chem. and Ind. *1963,* 53
5. Sargeant, K., Sheridan, A., O'Kelly, J., Carnaghan, R. B. A.: Nature *192,* 1096 (1961)
6. Spensley, P. C.: Endeavour *22,* 75 (1963)
7. Consumer Reports *1978,* 437
8. Patterson, D. S. P.: Cah. Nut. Diet. *11,* 2nd supplement, 71 (1976)
9. Stubblefield, R. D., Shotwell, O. L., Shannon, G. M.: J. Assoc. Off. Anal. Chem. *55,* 762 (1972)
10. Stubblefield, R. D., Shotwell, O. L., Shannon, G. M., Weisleder, D., Rohwedder, W. K.: J. Agric. Food Chem. *18,* 391 (1970)
11. Symposium on the Control of Mycotoxins, 1972: Pure and Appl. Chem. *35,* 209 (1973)
12. Third International IUPAC Symposium on Mycotoxins in Foodstuffs, 1976: Pure and Appl. Chem. *49,* 1703 (1977)
13. Wong, J. J., Hsieh, D. P. H.: Proc. Natl. Acad. Sci. U. S. A. *73,* 2241 (1976)
14. Wogan, G. N., Pong, R. S.: Ann. N. Y. Acad. Sci. *174,* 623 (1970)
15. Schoental, R.: A. Rev. Pharmacol. *7,* 343 (1967)
16. Kraybill, H. F.: Trop. Geogr. Med. *21,* 1 (1969)
17. Lilly, L. J.: Nature *207,* 433 (1965)
18. Butler, W. H., Clifford, J. I.: Nature *206,* 1045 (1965)
19. Hartley, R. D., Nesbitt, B. F., O'Kelly, J.: Nature *198,* 1056 (1963)
20. Dickens, F., Jones, H. E. H.: Brit. J. Cancer *19,* 392 (1965)
21. Butler, W. H.: Brit. J. Cancer *18,* 756 (1964)
22. Rogers, A. E.: Reyes' Syndr. Proc. Conf. (M. I. T.), 135–145, 421–453 (1974)
23. Purchase, I. F. H.: Food Cosmet. Toxicol. *5,* 339 (1967)
24. Pong, R. S., Wogan, G. N.: J. Nat. Cancer Inst. *47,* 585 (1971)
25. Wogan, G. N., Edwards, G. S., Newberne, P. M.: Cancer Res. *31,* 1936 (1971)
26. Gorst-Allman, C. P., Steyn, P. S., Wessels, P. L.: J. Chem. Soc. Perkin I *1977,* 1360
27. Wogan, G. N.: Trout Hepatoma Res. Conf. Papers, Res. Rep. No. *70,* 121 (1967)
28. Buchi, G.: Plenary Main Sect. Lect., Int. Cong. Pure Appl. Chem., 24th: *2,* 87 (1973): Published 1974
29. Goldblatt, L. A.: Pure and Appl. Chem. *21,* 331 (1970)
30. Fishbein, L., Falk, H. L.: Chromatographic Rev. *12,* 42 (1970)
31. Maggon, K. K., Viswanathan, L., Venkitasubramanian: J. Scient. Ind. Res. *29,* 8 (1970)
32. Aflatoxin. by Goldblatt, L. A.: New York: Academic Press 1969
33. Butler, W. H.: Mycotoxins *1974,* 1428
34. Wogan, G. N.: liver *1973,* 161
35. Pokrovskii, A. A., Lashneva, N. V., Staneva, M. P., Kravchenko, L. V., Valdes-Mendoza, V. S., Smirnova, L. A., Nikolaeva, M.: Probl. Med. Khim. *1973,* 106
36. Jones, B. D.: Mycotoxic Fungi, Mycotoxins, Mycotoxicoses *1,* 201 (1977)
37. Teichmann, G., Krug, G.: Z. Chem. *14,* 444 (1974)
38. Seitz, L. M.: J. Chromatogr. *104,* 81 (1975)
39. Pons, W. A.: J. Assoc. Off. Anal. Chem. *59,* 101 (1976)
40. Stoloff, L.: Mycotoxins, Hum. Anim. Health Proc. Conf. *1977,* 7
41. Pons, W. A., Robertson, J. A., Goldblatt, L. A.: J. Amer. Oil Chem. Soc. *43,* 665 (1966)
42. Andrellos, P. J., Beckwith, A. C., Eppley, R. M.: J. Assoc. Off. Anal. Chem. *50,* 346 (1967)
43. Ciegler, A., Lillehoj, A. B., Peterson, R. E., Hall, H. H.: Appl. Microbiol. *14,* 934 (1966)
44. Lee, L. S., Cucullu, A. F., Goldblatt, L. A.: Food Technol. *22,* 1131 (1968)
45. Trager, W., Stoloff, L.: J. Agr. Food Chem. *15,* 679 (1967)
46. Purchase, I. F. H., Steyn, M., Risma, R., Tustin, R. C.: Food Cosmet. Toxicol *10,* 383 (1972)
47. Goded, M. A.: Ion *36,* 367 (1976)
48. Goldblatt, L. A.: J. Amer. Oil Chem. Soc. *54,* 302A (1977)
49. Bergot, B. J., Stanley, N. L., Masri, M. S.: J. Agric. Food Chem. *25,* 965 (1977)
50. Yuichi, H., Hamasaki, T., Kagaku, T.: Seibutsu *11,* 552 (1973)
51. Nesbitt, B. F., O'Kelly, J., Sargeant, K., Sheridan, A.: Nature *195,* 1062 (1962)
52. Coomes, T. J., Cornelius, J. A., Shone, G.: Chem. and Ind. *1963,* 367
53. Hesseltine, C. W., Shotwell, O. L., Ellis, J. J., Stubblefield, R. D.: Bact. Rev. *30,* 795 (1966)

54. Asao, T., Buchi, G., Abdel-Kader, M. M., Chang, S. B., Wick, E. L., Wogan, G. N.: J. Amer. Chem. Soc. *85,* 1706 (1963)
55. Asao, T., Buchi, G., Abdel-Kader, M. M., Chang, S. B., Wick, E. L., Wogan, G. N.: J. Amer. Chem. Soc. *87,* 882 (1965)
56. Van der Merwe, K. J., Fourie, L., Scott, dB.: Chem. and Ind. *1963,* 1660
57. Van Dorp, D. A., van der Zijden, A. S. M., Beerthuis, R. K., Sparreboom, S., Ord, W. O., DeJong, K., Keunig, R.: Rec. Trav. Chim. *82,* 587 (1963)
58. Cheung, K. K., Sim, G. A.: Nature *201,* 1185 (1964)
59. Bullock, E., Roberts, J. C., Underwood, J. G.: J. Chem. Soc. *1962,* 4179
60. Bullock, E., Kirkaldy, D., Roberts, J. C., Underwood, J. G.: J. Chem. Soc. *1963,* 829
61. Van Soest, T. C., Peerdeman, A. F.: Konikl. Ned. Akad. Wetenschap. Proc. Ser. B *67,* 469 (1964)
62. Brechbuhler, S., Buchi, G., Milne, G.: J. Org. Chem. *32,* 2641 (1967)
63. Chang, S. B., Abdel-Kader, M. M., Wick, E. L., Wogan, G. N.: Science *142,* 1191 (1963)
64. Dutton, M. F., Heathcoate, J. G.: Chem. and Ind. *1968,* 418
65. Schabort, J. C., Steyn, M.: Biochem. Pharmacol. *21,* 2937 (1972)
66. Buchi, G., Foulkes, D. M., Kurono, M., Mitchell, G. F.: J. Amer. Chem. Soc. *88,* 4535 (1966)
67. Pohland, A. E., Cushmac, M. E., Andrellos, P. J.: J. Assoc. Off. Anal. Chem. *51,* 907 (1968)
68. Ciegler, A., Peterson, R. E.: Appl. Microbiol. *16,* 665 (1968)
69. Pohland, A. E., Yin, L., Dantzman, J. G.: J. Assoc. Off. Anal. Chem. *53,* 101 (1970)
70. Andrellos, P. J., Reid, J. R.: J. Assoc. Off. Anal. Chem. *47,* 801 (1964)
71. Allcroft, R., Carnaghan, R. B. A.: J. Vet. Rec. *74,* 863 (1962)
72. DeIongh, H., Vles, R. O., van Pelt, J. G.: Nature *202,* 466 (1964)
73. Schabort, J. C., Steyn, M.: Biochem. Pharmacol. *18,* 2441 (1969)
74. Allcroft, R., Rogers, H., Lewis, G., Nabney, J., Best, P. E.: Nature *209,* 154 (1966)
75. Branol, E.: Wien. Tierärztl. Monatsschr. *63,* 166 (1976)
76. Holzapfel, C. W., Steyn, P. S., Purchase, I. F. H.: Tetrahedron Lett. *1966,* 2799
77. Masri, M. S., Lundin, R. E., Page, J. R., Garcia, V. C.: Nature *215,* 753 (1967)
78. Nabney, J., Burbage, M. B., Allcroft, R., Lewis, G.: Food Cosmet. Toxicol. *5,* 11 (1967)
79. Purchase, I. F. H., Theron, J. J.: International Pathol. *8,* 3 (1967)
80. Heathcoate, J. G., Dutton, M. F.: Tetrahedron *25,* 1497 (1969)
81. Patterson, D. S. P., Roberts, B. A.: Food Cosmet. Toxicol. *9,* 929 (1971)
82. Detroy, R. W., Hesseltine, C. W.: Can. J. Biochem. *48,* 830 (1970)
83. Dalezios, J. I., Wogan, G. N., Weinreb, S. M.: Science *171,* 584 (1971)
84. Buchi, G., Spitzner, D., Palialunga, S., Wogan, G. N.: Life Sci. *13,* 1143 (1973)
85. Steyn, P. S., Vleggaar, R., Pitout, M. J., Steyn, M., Thiel, P. G.: J. Chem. Soc. Perkin I *1974,* 2551
86. Masri, M. S., Haddon, W. F., Lundin, R. E., Hsieh, D. P. H.: J. Agric. Food Chem. *22,* 512 (1974)
87. Ashoor, S., Chu, F. S.: J. Assoc. Off. Anal. Chem. *58,* 492 (1975)
88. Jones, B. D.: Mycotoxic Fungi, Mycotoxins, Mycotoxicoses *1,* 136 (1977)
89. Goldblatt, L. A., Dolliar, F. G.: Pure and Appl. Chem. *49,* 1759 (1977)
90. Cucullu, A. F., Lee, L. S., Pons, W. A., Stanley, J. B.: J. Agric. Food Chem. *24,* 408 (1976)
91. Hatsuda, Y., and Kuyama, S., Terashima, N.: J. Agric. Chem. Soc. Jap. *28,* 992 (1954)
92. Birkinshaw, J. H., Hammady, I. M. M.: Biochem. J. *65,* 162 (1957)
93. Davies, J. E., Roberts, J. C., Wallwork, S. C.: Chem. and Ind. *1956,* 178
94. Holzapfel, C. W., Purchase, I. F. H., Steyn, P. S., Gouws, L.: South African Med. J. *40,* 1100 (1966)
95. Davies, J. E., Kirkaldy, D., Roberts, J. C.: J. Chem. Soc. *1960,* 2169
96. Roberts, J. C., Underwood, J. G.: J. Chem. Soc. *1962,* 2060
97. Roberts, J. C.: J. Chem. Soc. *1960,* 785
98. Hatsuda, Y., Kuyama, S.: J. Agric. Chem. Soc. Jap. *28,* 989 (1954)
99. Roberts, J. C.: Chem. Rev. *61,* 591 (1961)
100. Tanaka, N., Katsube, Y., Hatsuda, Y., Hamasaki, T., Ishida, M.: Bull. Chem. Soc. Jap. *11,* 3635 (1970)

101. Holker, J. S. E., Mulheirn, L. J.: J. Chem. Soc. Chem. Commun. *1968,* 1576
102. Burkhardt, H. J., Forgacs, J.: Tetrahedron *24,* 717 (1968)
103. Elsworthy, G. C., Holker, J. S. E., McKeown, J. M., Robinson, J. B., Mulheirn, L. J.: J. Chem. Soc. Chem. Commun. *1970,* 1069
104. Groger, D., Erge, D., Franck, B., Ohnsorge, U., Flasch, H., Huber, F.: Chem. Ber. *101,* 1970 (1968)
105. Cole, R. J., Kirksey, J. W., Schroeder, H. W.: Tetrahedron Lett. *1970,* 3109
106. Holker, J. S. E., Kagal, S. A.: J. Chem. Soc. Chem. Commun. *1968,* 1574
107. Rodricks, J. V., Henery-Logan, K. R., Campbell, A. D., Stoloff, L., Verrett, M. J.: Nature *217,* 668 (1968)
108. Rodricks, J. V., Lustig, E., Campbell, A. D., Stoloff, L.: Tetrahedron Lett. *1968,* 2975
109. Waiss, A. C., Wiley, M., Black, D. R., Lundin, R. E.: Tetrahedron Lett. *1968,* 3207
110. Vleggaar, R., Steyn, P. S., Nagel, D. W.: J. Chem. Soc. Perkin I *1974,* 45
111. Steyn, P. S., Vleggaar, R.: J. Chem. Soc. Perkin I *1974,* 2250
112. Hamasaki, T., Hatsuda, Y., Terashima, N., Renbutsu, M.: Agric. and Biol. Chem. *31,* 11 (1967)
113. Lee, L. S., Bennett, J. W., Cucullu, A. F., Stanley, J. B.: J. Agric. Food Chem. *23,* 1132 (1975)
114. Buchi, G., Rae, I. D.: The Structure and Chemistry of the Aflatoxins. In: Aflatoxin. Goldblatt, L. A., ed., London – New York: Academic Press 1969
115. Dean, F. M.: The Total Synthesis of Naturally Occuring Oxygen Ring Compounds- Sterigmatocystin and the Aflatoxins. In: The Total Synthesis of Natural Products, Vol. 1. Ap-Simon, J., ed., pp. 485–498. New York: Wiley Interscience 1973
116. Roberts, J. C.: Fortschr. Chem. Org. Naturst. *31,* 119 (1974)
117. Knight, J. A., Roberts, J. C., Roffey, P.: J. Chem. Soc. C *1966,* 1308
118. von Pechmann, H.: Chem. Ber. *17,* 929 (1884)
119. von Pechmann, H., Duisberg, C.: Chem. Ber. *16,* 2119 (1883)
120. Barris, E., Israelstam, S. S.: Chem. and Ind. *1958,* 1430
121. Desai, R. D., Ahmad, S. Z.: Proc. Ind. Acad. Sci. *5,* 277 (1937)
122. John, E. V. O., Israelstam, S. S.: J. Org. Chem. *26,* 240 (1961)
123. Desai, R. D., Ahmad, S. Z.: Proc. Ind. Acad. Sci. *6,* 6 (1938)
124. Wawzonek, S.: Coumarins Chapter Six. In: Heterocyclic Compounds. Elderfield, R. C., ed., New York: Wiley 1951
125. Winkler, D. E., Whetstone, R. R.: J. Org. Chem. *26,* 784 (1961)
126. Cook, C. E., Corley, R. C., Wall, M. E.: J. Org. Chem. *30,* 4114 (1965)
127. Kappe, T., Ziegler, E.: Org. Prep. and Proc. *1,* 61 (1969)
128. Thakor, V. M., Acharya, A. B.: J. Ind. Chem. Soc. *48,* 12 (1971)
129. Chatterji, D., Chakraborty, D. P.: J. Ind. Chem. Soc. *49,* 1045 (1972), and references cited therein
130. Sethna, S. M., Shah, N. M.: Chem. Rev. *36,* 1 (1945)
131. Sethna, S. M., Phadke, R.: Org. React. *7,* 1 (1945)
132. Holker, J. S. E., Underwood, J. G.: Chem. and Ind. *1964,* 1865
133. Pratt, D. D., Robinson, R.: J. Chem. Soc. *1924,* 188
134. Banerjee, D. K., Sivanandaian, K. M.: J. Org. Chem. *26,* 1634 (1961), and references cited therein
135. Taylor, E. C., McKillop, A.: Tetrahedron *23,* 897 (1967)
136. Buchi, G., Foulkes, D. M., Kurono, M., Mitchell, G. F., Schneider, R. S.: J. Amer. Chem. Soc. *89,* 6745 (1967)
137. Sonn, A., Bulow, W.: Chem. Ber. *58,* 1691 (1925)
138. Sawhney, P. L., Seshadri, T. R.: Proc. Ind. Acad. Sci. *37A,* 592 (1953)
139. Buchi, G., Weinreb, S. M.: J. Amer. Chem. Soc. *93,* 746 (1971)
140. Buchi, G., Roberts, E. C.: J. Org. Chem. *33,* 460 (1968)
141. Knight, J. A., Roberts, J. C., Roffey, P., Sheppard, A. H.: J. Chem. Soc. Chem. Commun. *1966,* 706
142. Roberts, J. C., Sheppard, A. H., Knight, J. A., Roffey, P.: J. Chem. Soc. C *1968,* 22

143. Weiland, J. H. S.: Rev. Trav. Chim. *83,* 81 (1964)
144. Bycroft, B. W., Hatton, J. R., Roberts, J. C.: J. Chem. Soc. C *1970,* 281
145. Buchi, G., Weinreb, S. M.: J. Amer. Chem. Soc. *91,* 5408 (1969)
146. Geissman, T. A., Hinreiner, E.: J. Amer. Chem. Soc. *73,* 782 (1951)
147. Mulholland, T. P. C., Ward, G.: J. Chem. Soc.*1953,* 1642
148. Duncanson, L. A., Grove, J. F., MacMillan, J., Mulholland, T. P. C.: J. Chem. Soc. *1957,* 3555
149. Klaubert, D.: PhD Thesis; Massachusetts Institute of Technology, 1971
150. Buchi, G., Luk, K. C., Muller, P. M.: J. Org. Chem. *40,* 3458 (1975)
151. Rance, M. J., Roberts, J. C.: Tetrahedron Lett. *1969,* 277
152. Williams, A. L., Kinney, R. E., Bridger, R. F.: J. Org. Chem. *32,* 2501 (1967)
153. Holmwood, G. M., Roberts, J. C.: Tetrahedron Lett. *1971,* 833
154. Adams, R., Mackenzie, S., Loewe, S.: J. Amer. Chem. Soc. *70,* 664 (1948)
155. Adams, R., Harfenist, M., Loewe, S.: J. Amer. Chem. Soc. *71,* 1624 (1949)
156. Davies, J. S., Davies, V. H., Hassall, C. H.: J. Chem. Soc. C *1969,* 1874
157. Hassall, C. H., Morgan, B. A.: J. Chem. Soc. Chem. Commun. *1970,* 1345
158. Rance, M. J., Roberts, J. C.: Tetrahedron Lett. *1970,* 2799
159. Gronowitz, S., Sorlin, J.: Arkiv fur Kemi *19,* 515 (1962)
160. Fisher, B. E., Hodge, J. E.: J. Org. Chem. *29,* 776 (1964)
161. Hofmann, A., Philipsborn, W., Eugster, C. H.: Helv. Chim. Acta. *48,* 1322 (1965)
162. Buchi, G., Liesch, J., Schuda, P. F.: Private communications and as yet unpublished results
163. Kampouris, E. M.: J. Chem. Soc. *1965,* 2651
164. Kampouris, E. M.: J. Chem. Soc. C *1968,* 2125
165. Czernecki, S., Georgoulis, C., Provelenghiou, C.: Tetrahedron Lett. *1976,* 3535
166. Otey, F. H., Mehltretter, C. L.: J. Org. Chem. *26,* 1673 (1961)
167. Serebrennikova, G. A., Vtorov, I. B., Preobrazhenskii, N. A.: Zhur. Org. Khim. *5,* 676 (1969); English Translation: p. 663 (1969)
168. Corey, E. J., Suggs, J. W.: J. Org. Chem. *38,* 3224 (1973)
169. For fairly recent summaries, see Refs. 16 and 31, as well as the references cited therein
170. Thomas, R.: Biogenesis of Antibiotic Substances. Vanek, Z. and Hestalek, Z., ed. p. 155. London – New York: Academic Press 1965
171. Lin, M. T., Hsieh, D. P. H.: J. Amer. Chem. Soc. *95,* 1668 (1973)
172. Hsieh, D. P. H., Yao, R. C., Fitzell, D. L., Reece, C. A.: J. Amer. Chem. Soc. *98,* 1020 (1976)
173. Hsieh, D. P. H., Seiber, J. N., Reece, C. A., Fitzell, D. L., Yang, S. L., Dalezios, J. I., LaMar, G. N., Budd, D. L., Motell, E.: Tetrahedron *31,* 661 (1975)
174. Steyn, P. S., Vleggaar, R., Wessels, P. L.: J. Chem. Soc. Chem. Commun. *1975,* 193
175. Kingston, D. G. I., Chen, P. N., Vercelloti: Phytochem. *15,* 1037 (1976)
176. Tanabe, M., Uramoto, M., Hamasaki, T., Cary, L.: Heterocycles *5,* 355 (1976)
177. Thomas, R.: personal communication to Moss, M. O. In: Phytochem. Ecology. Harborne, J. B., ed., p. 140. London – New York: Academic Press 1972
178. Hsieh, D. P. H., Lin, M. T., Yao, R. C., Singh, R.: J. Agric. Food Chem. *24,* 1170 (1976)
179. Yao, R. C., Hsieh, D. P. H.: Appl. Microbiol. *28,* 52 (1974)
180. Lee, L. S., Bennett, J. W., Cucullu, A. F., Ory, R. L.: J. Agric. Food Chem. *24,* 1167 (1976)
181. Hsieh, D. P. H., Lin, M. T., Yao, R. C.: Biochem. Biophys. Res. Commun. *47,* 1051 (1972)
182. Hsieh, D. P. H., Lin, M. T., Yao, R. C.: Biochem. Biophys. Res. Commun. *52,* 992 (1973)
183. Fitzell, D. L., Hsieh, D. P. H., Yao, R. C., LaMar, J. N.: J. Agric. Food. Chem. *23,* 442 (1975)
184. Pachler, K. G. R., Steyn, P. S., Vleggaar, R., Wessels, P. L.: J. Chem. Soc. Perkin I *1976,* 1182; and references cited therein
185. Cox, R. H., and Cole, R. J.: J. Org. Chem. *42,* 112 (1977)

186. Gorst-Allman, C. P., Pachler, K. G. R., Steyn, P. S., Wessels, P. L.: J. Chem. Soc. Perkin I *1977*, 2181; and references cited therein
187. Gorst-Allman, C. P., Steyn, P. S., Wessels, P. L.: J. Chem. Soc. Perkin I 1978, 961; and references cited therein
188. Essigman, J. M., Croy, R. G., Nadzan, A. M., Busby, W. F., Reinhold, V. N., Buchi, G., Wogan, G. N.: Science *74*, 1870 (1977)

Received April 17, 1979

Author Index Volumes 50–91

The volume numbers are printed in italics

Reactivity and Structure

Concepts in Organic Chemistry

Editors: K. Hafner, J.-M. Lehn, C. W. Rees, P. v. Ragué Schleyer, B. M. Trost, R. Zahradnik

This series will not only deal with problems of the reactivity and structure of organic compounds but also consider synthetical-preparative aspects.
Suggestions as to topics will always be welcome.

Volume 1: J. Tsuji
Organic Synthesis
by Means of Transition Metal Complexes
A Systematic Approach
1975. 4 tables. IX, 199 pages
ISBN 3-540-07227-6

Volume 2: K. Fukui
Theory of Orientation and Stereoselection
1975. 72 figures, 2 tables. VII, 134 pages
ISBN 3-540-07426-0

Volume 3: H. Kwart, K. King
d-Orbitals in the Chemistry of Silicon, Phosphorus and Sulfur
1977. 4 figures, 10 tables. VIII, 220 pages
ISBN 3-540-07953-X

Volume 4: W. P. Weber, G. W. Gokel
Phase Transfer Catalysis in Organic Synthesis
1977. 100 tables. XV, 280 pages
ISBN 3-540-08377-4

Volume 5: N. D. Epiotis
Theory of Organic Reactions
1978. 69 figures, 47 tables. XIV, 290 pages
ISBN 3-540-08551-3

Volume 6: M. L. Bender, M. Komiyama
Cyclodextrin Chemistry
1978. 14 figures, 37 tables. X, 96 pages
ISBN 3-540-08577-7

Volume 7: D. I. Davies, M. J. Parrott
Free Radicals in Organic Synthesis
1978. 1 figure. XII, 169 pages
ISBN 3-540-08723-0

Volume 8: C. Birr
Aspects of the Merrifield Peptide Synthesis
1978. 62 figures, 6 tables. VIII, 102 pages
ISBN 3-540-08872-5

Volume 9: J. R. Blackborow, D. Young
Metal Vapour Synthesis in Organometallic Chemistry
1979. 36 figures, 32 tables. XIII, 202 pages
ISBN 3-540-09330-3

Volume 10: J. Tsuji
Organic Synthesis with Palladium Compounds
1980. 9 tables. XII, 207 pages
ISBN 3-540-09767-8

Volume 11:
New Syntheses with Carbon Monoxide
Editor: J. Falbe
1980. 118 figures, 127 tables.
Approx. 490 pages
ISBN 3-540-09674-4

Volume 12: J. Fabian, H. Hartmann
Light Absorption of Organic Colorants
Theoretical Treatment and Empirical Rules
1980. 76 figures, 68 tables. Approx. 260 pages
ISBN 3-540-09914-X

**Springer-Verlag
Berlin Heidelberg NewYork**

Topics in Current Chemistry

Fortschritte der chemischen Forschung

Managing Editor: F. L. Boschke

A Selection

Volume 83

Biochemistry

1979. 84 figures, 20 tables. IV, 178 pages
ISBN 3-540-09312-5

Contents:

H. F. DeLuca, H. E. Paaren, H. K. Schnoes: Vitamin D and Calcium Metabolism (413 ref.)
V. Ullrich: Cytochrome P450 and Biological Hydroxylation Reactions (166 ref.). *J. Reden, W. Dürckheimer:* Aminoglycoside Antibiotics – Chemistry, Biochemistry, Structure-Activity Relationships (309 ref.).

Volume 84

Bioactive Organo-Silicon Compounds

1979. 21 figures, 50 tables. IV, 146 pages
ISBN 3-540-09347-8

Contents:

R. Tacke, U. Wannagat: Syntheses and Properties of Bioactive Organo-Silicon Compounds (132 ref.).
M. G. Voronkov: Biological Activity of Silatranes (101 ref.).

Volume 85

Instrumental Inorganic Chemistry

1979. 64 figures, 23 tables. IV, 220 pages
ISBN 3-540-09338-9

Contents:

G. K. Wolf: Chemical Effects of Ion Bombardment (289 ref.). *R. W. Kiser:* Doubly-Charged Negative Ions in the Gas Phase (318 ref.). *G. Schwedt:* Chromatography in Inorganic Trace Analysis (345 ref.)

Volume 86

Spectroscopy

1979. 136 figures, 22 tables. IV, 294 pages
ISBN 3-540-09462-8

Contents:

M. N. R. Ashfold, M. T. Macpherson, J. P. Simons: Photochemistry and Spectroscopy of Simple Polyatomic Molecules in the Vacuum Ultraviolet (365 ref.). *C. Sandorfy:* Far-Ultraviolet Absorption Spectra of Organic Molecules: Valence-Shell and Rydberg Transitions (180 ref.). *R. Gleiter, J. Spanget-Larsen:* Some Aspects of the Photoelectron Spectroscopy of Organic Sulfur Compounds (132 ref.). *R. Gleiter:* Photoelectron Spectra and Bonding in Small Ring Hydrocarbons (127 ref.)

Volume 87

Micelles

1980. 69 figures, 12 tables. IV, 154 pages
ISBN 3-540-09639-6

Contents:

B. Lindman, H. Wennerström: Micelles/Amphiphile Aggregation in Aqueous Solution (317 ref.)
H.-F. Eicke: Surfactants in Nonpolar Solvents/ Aggregation and Micellization (228 ref.)

Volume 88

Organic Chemistry/ Syntheses and Reactivity

1980. 19 figures, 36 tables. IV, 170 pages
ISBN 3-540-09817-8

Contents:

C. Rüchardt: Steric Effects in Free Radical Chemistry (120 ref.). *L. Birkhofer, O. Stuhl:* Sylylated Synthons – Facile Organic Reagents of Great Applicability (323 ref.). *K. A. Muszkat:* The 4a, 4b-Dihydrophenanthrenes (90 ref.). *N. T. Anh:* Regio- and Stereo-Selectivities in Some Nucleophilic Reactions (46 ref.).

Volume 89

Plasma Chemistry I

1980. 52 figures, 5 tables. VI, 146 pages
ISBN 3-540-09825-9

Contents:

D. Smith, N. G. Adams: Elementary Plasma Reactions of Environmental Interest (214 ref.). *D. M. Gruen, S. Vepřek, R. B. Wright:* Plasma-Materials Interactions and Impurity Control in Magnetically Confined Thermonuclear Fusion Machines (207 ref.). *D. Küppers, H. Lydtin:* Preparation of Optical Waveguides with the Aid of Plasma-Activated Chemical Vapour Deposition at Low Pressures (24 ref.)

Volume 90

Plasma Chemistry II

1980. 67 figures, 25 tables. IV, 121 pages
ISBN 3-540-09826-7

Contents:

M. Venugopalan, U. K. Roychowdhury, K. Chan, M. L. Pool: Plasma Chemistry of Fossil Fuels (313 ref.). *M. Capitelli, E. Molinari:* Kinetics of Dissociation Processes in Plasmas in the Low and Intermediate Pressure Range (71 ref.)

Springer-Verlag
Berlin
Heidelberg
New York